T0331177

Distributed Artificial Intelligence for 5G/6G Communications

Distributed Artificial Intelligence for 5G/6G Communications

Frameworks with Machine Learning

Iacovos Ioannou, Prabagarane Nagaradjane,
Vasos Vassiliou, Christophoros Christophorou,
and Andreas Pitsillides

CRC Press is an imprint of the
Taylor & Francis Group, an informa business

First edition published 2025
by CRC Press
2385 NW Executive Center Drive, Suite 320, Boca Raton FL 33431

and by CRC Press
4 Park Square, Milton Park, Abingdon, Oxon, OX14 4RN

CRC Press is an imprint of Taylor & Francis Group, LLC

ISBN: 978-1-032-74435-3 (hbk)
ISBN: 978-1-032-74436-0 (pbk)
ISBN: 978-1-003-46920-9 (ebk)

DOI: 10.1201/9781003469209

Typeset in Times
by codeMantra

Contents

List of Figures

List of Tables

Preface

New generation networks, such as 5G and forthcoming 6G, face many technical challenges in reaching the very ambitious standards set forth by the research and the industrial community. These technical challenges include: (i) support for a very large number of devices under the same network; (ii) providing an ultra-reliable low latency communication; (iii) being dynamic and adaptable; and (iv) providing high service quality and quantity in terms of bandwidth. Given the above challenges, specific issues related to network management and control arise, including efficient communication establishment control, and a fast decision and disaster recovery. A joint management and control approach becomes necessary to handle the above issues effectively, with autonomous and adaptable actions.

In this book, inspired by the expected benefits of adopting artificial intelligence (AI) and machine learning (ML) approaches in 5G and 6G networks, we propose and develop a novel distributed AI (DAI) framework with AI/ML able to facilitate the achievement of the ambitious goals set forth. The proposed DAI framework utilizes Belief Desire Intention (BDI) agents extended with ML capabilities. We refer to these as BDIx agents. The BDIx agents reside on the mobile devices, forming a multi-agent system (MAS) integrating fuzzy logic and back-propagation neural network for reinforcement learning at the perception/cognitive part of the agents.

To illustrate the potential of the DAI framework, we focus on the aspect of Device-to-Device (D2D) communication in 5G and beyond networks. Its inherently distributed nature with a vast number of user devices/user equipment (UEs) makes it appealing for the application and demonstration of the DAI framework, incorporating BDIx agents in the D2D UEs. The main advantage of D2D communication is that the licensed frequency bands do not constrain it, and it is transparent to the cellular network. It permits adjacent UEs to bypass the Base Station (BS) and establish direct links between them. By enabling this, improved spectral efficiency, energy efficiency, data rates, throughput, delay, interference, and fairness can be achieved. The above-noted improvements in network performance spearheaded a vast amount of research in D2D, which identified significant challenges to be addressed before realizing their full potential in 5G and 6G. The DAI framework is expected to be a supporting pillar in addressing these challenges.

Furthermore, through the specific example of mode selection in D2D 5G, we design and develop a detailed DAI solution (DAIS) framework plan, discuss implementation complexities and technology aspects, and then implement the DAIS algorithm/plan, executed by the BDIx agents at a static and dynamic network with speed and direction. We demonstrate its benefits, such as the BDIx agent's capabilities in intercommunication and cooperation in an efficient, distributed, autonomous and flexible manner, thus offering improved performance. Extensive simulative evaluations using representative metrics (spectral efficiency and power consumption), the well-known quality of service and customer satisfaction metrics (QoS and QoE), custom-made metrics (D2D effectiveness, stability, and productivity metrics), and specific metrics (cluster formation, message exchange, and control decision delay), are carried out. Additionally, a comparative evaluation is performed in a static environment against distributed sum rate (DSR) with global knowledge, as well as potentially competing techniques, such as Fuzzy Adaptive Resonance Theory (Fuzzy ART), Density-Based Scan (DBSCAN), Gaussian expectation-maximization (G-MEANS), and Minimum Entropy Clustering (MEC), customized to the needs of D2D communication. Moreover, a comparative evaluation is performed in a dynamic environment with speed and direction against the distributed sum rate (DSR) approach with global knowledge and potentially competing techniques, such as enhanced single hop relay (SHRA). Important lessons learned are discussed and analyzed in the book as well as in future work.

This book demonstrates that the DAI framework can offer fast network control with less messaging exchange, reduced signalling overhead, and fast decision-making. Also, it can support self-healing mechanisms and act collaboratively as a self-organizing network. Additionally, it can capitalize on existing implementations, e.g., artificial neural networks, for tackling any other D2D challenges or any other 5G and 6G challenges.

Iacovos Ioannou

Acknowledgements

This book is largely based on the PhD Thesis of Iacovos Ioannou, submitted to the Computer Science Department of the University of Cyprus (https://gnosis.library.ucy.ac.cy/handle/7/65407). The following text forms an acknowledgment of the core people who helped Dr Ioannou while pursuing a PhD degree. "First and foremost, I am incredibly grateful to my supervisors, Professor Andreas Pitsillides and Associate Professor Vasos Vassiliou, for their invaluable advice, continuous support, and patience during my PhD study. Their immense knowledge and experience have always guided me in my academic research and daily life. Also, I would like to thank Dr Christophoros Christophorou for his valuable scientific help and mental support of my study, which made the achievement of the PhD an exciting journey. Additionally, I would like to thank Dr Christiana Ioannou for her help and mental support of my study. Finally, I would like to express my gratitude to my wife, Maria K. Ioannou, my daughter Andri Ioannou, and my grandfather, Socrates, for their precious advice, and last but not least, my parents, Ioannis and Paraskevi Ioannou. Without their tremendous understanding and encouragement in the past few years, completing my study would have been impossible." This work was supported by the ADROIT6G Project, which is funded by the Smart Networks and Services Joint Undertaking (SNS JU) under the European Union's Horizon Europe research and innovation program (Grant Agreement No. 101095363). The project aims to establish a cutting-edge wireless system architecture for 6G that integrates distributed computing nodes at the far-edge, edge, and cloud domains. We also wish to thank my research center CYENS, and the collaborators of the ADROIT6G project such as CNIT, EURE, OULO, Nvidia, NOVA, SIEMENS, ORANGE, CAFA-TECH, EBOS, IQUADRAT, and NEXTWORKS for their contributions in advancing research in distributed and dynamic network systems, as well as the European Commission for supporting this initiative.

About the Authors

Dr. Iacovos Ioannou is a researcher affiliated with CYENS and the Networks Research Laboratory at the University of Cyprus. His research focuses on emerging fields including mobile and Internet of Things (IoT) network security, 5G and 6G wireless communications, and the incorporation of distributed artificial intelligence (DAI), artificial intelligence (AI), and machine learning (ML) in device-to-device (D2D) communications. His areas of competence encompass a wide range of activities, including the analysis, development, installation, and maintenance of IT systems and applications.

Dr. Vasos Vassiliou is an Associate Professor with the Computer Science Department of the University of Cyprus and the co-director of the Networks Research Laboratory (NetRL) at UCY (founded by Prof. A. Pitsillides). He is also the Group Leader of the Smart Networked Systems Research Group of the RISE Center of Excellence on Interactive Media, Smart Systems and Emerging Technologies, situated in Nicosia, Cyprus. Dr. Vassiliou has been appointed by the Senate of the University of Cyprus to the Board of Directors of CYNET, the National Research and Educational Network, where he has served as the Chair since 2016.

Dr. Prabagarane Nagaradjane is an Associate Professor with the Department of Electronics and Communication Engineering, SSN Institutions. He received M.Tech and Ph.D. in ECE from Pondicherry Engineering College, Pondicherry (Central) University. Before this, he was with HCL Info Systems for a brief period of two years. His research interests include various aspects of wireless communications, especially concerning signal processing for wireless and broadband communications and AI/ML applications in wireless communications.

Dr. Christophoros Christophorou completed his undergraduate and graduate studies at the University of Cyprus (B.Sc. in Computer Science and M.Sc. in Advanced Computer Technologies, 2002 and 2005). He received his PhD, also at the University of Cyprus, in the area of Mobile/Wireless Networks in 2011. His main research interests and expertise are primarily concentrated in the Telecommunications and Networking research fields but also include database management systems, social collaborative care networks, and Information and Communication Technology (ICT) personalized solutions (for various sectors, including the eHealth domain).

Dr. Andreas Pitsillides is an Emeritus Professor in the Department of Computer Science, University of Cyprus and a Visiting Professor Department of Electrical & Electronic Engineering Science, University of Johannesburg. He was a co-director of the Networks Research Laboratory (UCY) and was appointed Visiting Professor at the University of the Witwatersrand (Wits), School of Electrical and Information Engineering. His broad research interests include communication networks, including programmable wireless environments enabled by software-defined metamaterials (reconfigurable intelligent surfaces), nanonetworks (bio and e-m), the Internet of Things, smart systems (e.g. smart grid), smart spaces (e.g. home, city), and e-health.

Important Acronyms Table

Abbreviation	Definition
ACID	Atomicity, Consistency, Isolation, Durabilit
ACO	Ant Colony Optimization
AI	Artificial Intelligence
ANN	Artificial Neural Network
AP	Access Point
API	Application Programming Interface
AR	Augmented Reality
AWGN	Additive White Gaussian Noise
BDI	Belief Desire Intention
BDIx	Belief Desire Intention eXtented
BPL	Battery Power Level
BPNN	Back-Propagation Neural Network
BS	Base Station
CA	Certificate Authority
CDO	Cell Densification and Offloading
CH	Cluster Head
CI	Collective Intelligence
CPICH	Common Pilot Channel
CPU	Central Processing Unit
CQI	Channel Quality Indicator
CSI	Channel Signal Indicator
D2D	Device-to-Device
D2D Multi Hop Relay	Device to Device Multi Hop Relay
D2D Relay	Device to Device Relay
D2DMHR	Device to Device Multi Hop Relay
D2D-Relay	Device to Device Relay AND/OR Device to Device Multi Hop Relay
D2DSHR	Device to Device Single Hop Relay
DAI	Distributed Artificial Intelligence
DAIS	Distributed Artificial Intelligence Solution
DAIS	Distributed Artificial Intelligence Solution
DBSCAN	Density-Based Scan
DD	Device Discovery
DI	Dynamic Impelementation
DM	Data Mining
DNN	Deep Neural Network
DPS	Distributed Problem Solving
DR	Distributed Random
DSR	Distributed Sum Rate
EA	Evolutionary Algorithms
eMBB	enhanced Mobile Broadband
eNB	enhanced Node B
EPC	Evolved Packet Core
FB	Frequency Bands
FIPA	The Foundation for Intelligent Physical Agents

Abbreviation	Definition
FIPA-ACL	Foundation for Intelligent Physical Agents - Agent Communication Language
FL	Fuzzy Logic
FMS	Frequency Mode Selection
FT	Fault Tolerance
Fuzzy ART	Fuzzy Adaptive Resonance Theory
GA	Genetic Algorithms
G-MEANS	Gaussian expectation-maximization
H˙D2D	Handover D2D
HetNet	Heterogeneous Networks
HETNET	Heterogeneous network
HO	Handover
ICCID	Integrated Circuit Card Identification Number
IM	Interference Management
IMEI	International Mobile Equipment Identity
IMSI	International Mobile Subscriber Identity
IoT	Internet of Things
IP	Internet Protocol
ISO-OSI	International organization of Standardization – Open System Interconnection
ITU	International Telecommunication Union
LDR	Link Data Rate
LTE	Long-Term Evolution
LTE ProSe	Long Term Evolution Proximity Services
MA	Mobile Agent
MAS	Multi-Agent System
MEC	Minimum Entropy Clustering
MEC	Multi-access edge computing
ML	Machine Learning
MME	Mobility Management Entity
mMTC	massive Machine Type Communication
mmW	millimeter Wave
MS	Mode Selection
MS	Mode Selection
MSISDN	Mobile Station International Subscriber Director Number
NCU	Non-cooperative users
NFAPI	network Functional Application Platform Interface
NFV	Network Function Virtualisation
NN	Neural Networks
OFDMA	orthogonal frequency-division multiple access
PAI	Parallel AI
PC	Power Consumption
P-C	Power Control
PKI	Primary Key Indicators
PNF	Physical Network Functions
PPP	Poisson Point Process
PSO	Particle Swarm Optimization
PUSCH	Physical Uplink Shared Channel
QL	Q-Learning
QoE	Quality of Experience
QoS	Quality of Service
QoS˙P	QoS/Path Selection (Routing)
RAT	Radio Access Technologies
REST	Representational State Transfer

Abbreviation	Definition
RF	Radio Frequency
RL	Reinforcement Learning
RRA	Radio Resource Allocation
RRA	Radio Resource Allocation
RSA	Rivest–Shamir–Adleman
S	Security
SDN	Software Defined Network
SE	Spectral Efficiency
SHRA	Single Hop Relay Approach
SIM	Subscriber Identity Module
SINR	Signal To Interference Noise Ratio
SNR	Signal to Noice Ratio
SOAP	Simple Object Access Protocol
SON	self-organized network
SR	Sum Rate
SSL	Secure Sockets Layer
TB	Thompson sampling and Bayesian control
TMS	Transmission Mode Selection
TP	Transmission Power
TS	Time Step
UE	User Equipment
URLL	Ultra-reliable Low Latency
V2V	Vehicle-to-Vehicle
VANETS	Vehicular Ad-Hoc Networks
VNF	Virtual Network Function
VR	Virtual Reality
WDR	Weighted Data Rate
WSN	Wireless Sensor Networks

1 Introduction

New generation mobile networks, such as 5G and the forthcoming 6G, face many technical challenges in reaching the very ambitious standards set forth by the research and industrial community. Additionally, with the rapid developments achieved in artificial intelligence (AI) and machine learning (ML) throughout the last few years in terms of optimization and accuracy rate, the time has come for both worlds to come together. Therefore, currently, mobile communication and AI/ML jointly are used in approaches that target the 5G and 6G challenges.

In this book, inspired by the expected benefits of AI and ML approaches in 5G and 6G networks, we propose and develop a novel AI/ML-based Distributed AI (DAI) framework able to facilitate the achievement of the ambitious goals set for emerging 5G and 6G networks. The proposed DAI framework utilizes Belief Desire Intention (BDI) agents extended with ML capabilities. We refer to these as BDIx agents. The BDIx agents reside on mobile devices, forming a multi-agent system (MAS) integrating fuzzy logic for the perception and cognitive parts of the agents.

Thus, in this book, we deploy AI techniques for providing solutions in 5G and 6G mobile network management and control. Furthermore, we adopt the view that a distributed AI approach is well suited to handle the complexities of today's networks in an effective manner, providing responsive and robust control, hence becoming independent and autonomous systems.

1.1 MOTIVATION OF THE BOOK

The ambitious goals set for emerging 5G and 6G networks force the academic community to seek alternative ways in order to meet these and hence realize the demanded mobile network infrastructure, and they arise many technical challenges to achieve both of them: the emerging goals of 5G/6G networks and the demanded mobile network infrastructure [1,2].

These technical challenges include: (i) support for a very large number of devices (IoT included) under the same network (e.g. 1000s devices per square kilometer), called massive Machine Type Communications (mMTC) [3]; (ii) to provide an ultra reliable low latency communication (1 ms) for supporting new applications, such as remote medical operations, and new technologies, such as Augmented Reality (AR) and Virtual Reality (VR), called Ultra Reliable Low Latency Communications (URLL) [3]; (iii) to offer fast action to handle dynamic aspects; and (iv) to provide high service quality and quantity in terms of bandwidth, in order to achieve the users demanding bandwidth that come from mobile applications that use live video, high-quality images, voice and text (e.g. 1 Gbps per user), called enhanced Mobile Broadband (eMBB) [3].

Given the above challenges, specific issues related to network management and control arise, including efficient communication establishment control, fast decision-

DOI: 10.1201/9781003469209-1

making, and disaster recovery. Efficient communication establishment control is becoming increasingly complex with the new 5G/6G requirements. Different approaches are proposed in standards, including AI and softwarization. However, these approaches are not running on the UEs, thus are not distributed, parallel, or DAI and are focused on or dependedent on the BS. A fast decision will make a difference in the quality of the communication for the network client devices due to their dynamicity. In order to realize fast decisions, the devices should be autonomous and dynamic due to the resultant reduced volume of messaging. By reducing messaging exchange, the delays are also reduced. For disaster recovery, the need for flexible communication that can act independently is essential (self-healing/self-organized networks). To handle above issues effectively, a joint management and control approach become necessary.

Furthermore, it appears to be commonly accepted that AI and ML, among other technologies, are expected to play a crucial role in 5G/6G networks [1, 2, 4–11], as they will shape the future communication networks in designing and optimizing 5G/6G architectures and protocols. An indicator of the level of interest toward this direction is the building by the International Telecommunication Union (ITU) of an AI/ML Toolkit (see ITU-T Y.3173 Framework [12]) for evaluating intelligence levels of future networks, including the IMT-2020 and ITU-T Y.3170-series [13].

Also, the latest literature in 6G [4–10, 14, 15] specifies that connectivity demands of smart networks and requirements of near-future services can be only satisfied by a fully decentralized control with virtual resources [10]. Thus, future networks are expected to change from centralized control to distributed control and become independent and autonomous systems [8]. Furthermore, the use of AI and ML at the edge, by bringing intelligence from centralized computing facilities to every terminal in the network, is also mandated [1, 4–8, 10, 14, 15]. This, combined with unsupervised learning and inter-user inter-operator knowledge sharing, will promote real-time network decisions [6]. Additionally, AI, deep learning, and ML techniques will enable 6G to establish self-organization strategies, including self-learning, self-configuration, self-healing, and self-optimization of network resources at the terminal level (mobile devices) [4–10, 14, 15]. Furthermore, Collective Intelligence (CI), AI, and ML can jointly achieve 5G/6G communication [16] through agent collaboration.

Towards this end, aspired by the adoption of CI, AI, and ML approaches in 5G and 6G, in this work, we present a novel DAI framework, which we anticipate that it can facilitate in the achievement of the demanding requirements of 5G and beyond. Thus, in this book, we consider a D2D setup in 5G and beyond communication network requirements. In this setup, each D2D device, by controlling its cellular (i.e., LTE, 5G) and WiFi interfaces, aims to achieve D2D communication. The target is to tackle the following D2D challenges, by focusing on the local environment of D2D communication (i.e., the Weighted Data Rate (WDR) of the D2D path as shown in Section 6.1.2, the D2D devices' coordinates in proximity, etc.), rather than the global environment: (i) device discovery; (ii) mode selection; (iii) interference management; (iv) power control; (v) security concerns; (vi) radio resource allocation; (vii) cell densification and offloading; (viii) QoS & QoE (Path Selection & Routing); (ix) mmWave communication; (x) handover; and (xi) non-cooperative users. Additionally, relying only on local environment results in reduced signalling overhead and much faster control decision-making, targeting the achievement of the demanding requirements.

More specifically, our approach targets to implement a distributed, autonomous, dynamic, and flexible DAI framework that utilizes BDIx agents (with reinforcement learning because BDI agents from their architecture act with reinforcement learning and ML), where each BDIx agent will reside on each UE. The DAI framework offers the following advantages: (i) fast network control with less messaging exchange and reduced signalling overhead; (ii) fast decision-making; (iii) support of self-healing mechanisms and collaboratively act as a self-organizing network; and (iv) can capitalize on existing implementations (e.g., artificial neural networks [17]) for tackling any other D2D challenges. In this work, we only consider a (semi)static environment where each entering D2D device does not have a speed greater that 1.5 m/s (i.e., pedestrian speed). Extensions to higher mobility will be considered in future work.

In order to achieve the above advantages, the framework's architecture is envisioned to be modular and utilize the DAI concept. This aim is to provide the framework with the ability to act as a glue in the employment of more than one successful, optimized intelligent approach relying only on local knowledge in D2D (e.g., use deep neural networks to identify the best frequency that reduces interference to be used by an entering D2D device). Thus, the Beliefs and Desires can be substituted or added as modules (extra AI/ML models) targeting the achievement of a specific task/requirement in 5G D2D communication (e.g., high data rate). Also, with the use of the BDIx agents in the framework, it achieves intercommunication and collaborative decisions with the use of messages[1] among them (i.e., propose, notify, and inform).

Additionally, in the existing literature, most D2D intelligent approaches have the following open issues: (i) lack of participation in the implementation of D2D challenges; (ii) not a lot of approaches are dynamic[2] and flexible[3]; (iii) opportunities of research in D2D challenges by other AI techniques; (iv) opportunities of research in security; (v) an intelligent autonomous[4] solution without the use of the global network data does not exist; (vi) a DAI implementation of intelligent approaches is lacking; (vii) no work that supports self-organizing networks in D2D exists; (viii) the papers identified in the literature about D2D promote hardware change at BS and UEs, which is an expensive and difficult task to do; (ix) even though D2D is a locality issue (i.e., only between the proximate D2D devices) most of the approaches handle it as a global issue (at the BS); (x) an intelligent approach utilizing all spectrum utilization methods is lacking; (xi) an intelligent approach utilizing all transmission modes is lacking.

1.2 BOOK CONTRIBUTION

The main book contribution is the motivation, design, and realization of a generic DAI framework, incorporating a special type of agent with Beliefs, Desires, and

Intentions (the BDI agent), which is extended with machine learning capabilities yielding the BDIx agent. This framework is expected to have wide applicability in mobile networks and effectively control the aforementioned management and control challenges. In this book, we adopt a D2D setting in 5G/6G to demonstrate the salient features of the DAI framework.

In particular,

- We performed a literature review on DAI, BDI agents, and AI/ML approaches in 5G D2D communication. With this review, we have identified the open issues and challenges in 5G D2D communication along with the need for a DAI framework.
- We motivate and design a DAI framework for tackling the demanding management and control challenges found in 5G and beyond mobile communication networks (see Section 1.1).
- We extend the BDI agent, adopted within the proposed DAI framework, with machine learning capabilities yielding the BDIx agent.
- We explore implementation aspects and realize the BDIx agents under the proposed DAI framework.
- We implement the architecture, Plan Library, execution flowchart and BDIx interpreter of the DAI framework with BDIx agents.
- We specify the BDIx agent settings, the potential metrics, the implementation constraints, and the implementation specifics for achieving D2D communication in the DAI framework.
- As an illustration of the generic nature of the DAI framework, we design a number of representative example plans within the DAI framework for tackling demanding D2D challenges.
- To better illustrate the use of the DAI framework, we focus on the mode selection of D2D communication in a static and dynamic environment and realize two BDIx agent plans, the first called DAIS and the second DSR.
- We extensively evaluate DAIS and DSR and also compare them with other AI/ML techniques (in some cases, suitably adapted to take advantage of the DAI framework).
- We achieve the maximization of the total spectral efficiency[5] (SE) (i.e., sum rate) and the reduction of the total power consumption (PC) of the existing mobile network infrastructure (non-D2D UE) using the DAIS algorithm as well as the other investigated unsupervised learning AI/ML clustering techniques under a base station (BS).
- We show that unsupervised learning techniques can be utilized in order to achieve equal or better results than the DAIS or DSR approach in terms of transmission mode selection.
- We show, according to each approach, the mean execution time that a D2D Device takes to conclude in the selection of transmission mode at the D2D communication network.
- We identify the existing open issues in D2D communication through the research and a literature review.

An overview of the contributions is given next.

The proposed DAI framework forms a multi-agent system (MAS) utilizing Belief Desire Intention (BDI) agents [18–21] extended with machine learning capabilities to address demanding 5G/6G challenges. We refer to these as BDIx agents. The BDIx agents reside on the mobile devices, allowing them to intercommunicate and cooperate in an efficient, distributed, autonomous, and flexible manner. For the perception/cognitive part of the agents, fuzzy logic is used in the book. It is worth mentioning here that we selected intelligent agents in our approach because of their ability to concurrently solve multiple complex problems [22]. Additionally, we investigated the main features of the framework and how the DAI framework is realized with the implementation of Beliefs, Desires Intentions Extended (BDIx) agents in a distributed and decentralized manner. We also examined the realization of the BDIx agent and its architecture along with the use of fuzzy logic as Plan Library. Furthermore, we show the implementation specifics of the DAI framework.

To demonstrate the potential of the DAI framework, we focus on the aspect of device-to-device (D2D) communication in 5G and beyond networks. Its inherently distributed nature, with a vast number of user devices (UEs) makes it appealing for the application and demonstration of the DAI framework, incorporating BDIx agents in the D2D UEs. The main advantage of D2D communication is that it is not constrained by the licensed frequency bands and also it is transparent to the cellular network. That is, it permits adjacent user equipments (UEs) to bypass the base station (BS) and establish direct links between them. By enabling this, improved spectral efficiency, energy efficiency, data rates, throughput, delay, interference and fairness [23–25] can be achieved. The above-noted improvements in network performance spearheaded a vast amount of research in D2D, which identified significant challenges (shown in Section 2.2.2) to be addressed before realizing their full potential in 5G and beyond networks. The DAI framework is expected to be a supporting pillar in addressing these challenges.

In order for D2D communication to succeed in a 5G and beyond network, it must address a number of D2D requirements/challenges (as discussed in Section 1.1). In the existing literature, we could not identify a DAI implementation in D2D, with almost all papers taking a global perspective, normally engaging the Base Station. Furthermore, not much support for self-organizing, autonomous networks in D2D was identified. This book shows how the DAI framework can achieve the D2D requirements with the use of Beliefs and Desires and a Plan Library with fuzzy logic. More precisely, D2D challenges are defined as requirements and indirectly as Desires with the purpose of the DAI framework BDIx agent to achieve them. Then, the D2D requirements are implemented as plans for intentions that are derived from the Desires of the D2D device. In addition, some D2D requirements must be handled during raised events (i.e. when a device is entering the D2D Network) or when some threshold values are violated. The relations between network events, BDIx agent's events, D2D challenges/requirements, and D2D Desires are defined in this book. Another important investigation carried out in this book is the relationship of D2D challenges between them and also indirectly among the Desires. More specifically, the definition of the two relations of dependency and association between D2D challenges is specified. Because, for dependency, some D2D challenges need other challenges to finish before they can be completed (e.g. Transmission mode must know the surroundings using Device Discovery), and when a D2D challenge along with other D2D challenges can be triggered at the same time and if both are highly related among them, these associations are also specified (e.g. the frequency selection mode and transmission mode selection).

Moreover, this book examines how Desires tackle the D2D challenges related to network events and thresholds through approaches/plans. More specifically, our investigation identifies the thresholds, events, and network events that are associated and codes their associations. The thresholds are creating events; these events then change or add intentions (from Beliefs), and through intentions, plans are executed. The plans and indirect intentions must have a specific order so that the BDIx agent can effectively achieve D2D communication in 5G and beyond. This book also investigates the cases of network events involving D2D challenges and indirectly Desires. For each case of network event, it associates it with the D2D challenges and indirectly with Desires and thresholds. The purpose of this is to restrict the deliberation in the agent and direct the change in priority of the Desires according to the achievement of the 5G D2D communication. Therefore, priority values are introduced and utilized in order to find a way to pre-specify the order of execution (with the use of fuzzy logic).

Continuing, a number of D2D challenges are provided as DAI framework plans, which can be used in each desire to tackle these D2D challenges/requirements. Also, the realization of BDIx agents in terms of existing programming frameworks is investigated in this book. Finally, a section on multi-agent systems and how, with the usage of game theory, the collaboration of the BDIx agents can be achieved in order to satisfy all the device users and the telecom operator is offered as future work.

Next, in order to demonstrate the potentials of the DAI framework in a static environment, a specific plan/solution developed for transmission mode selection, called DAIS, was proposed in this book. DAIS is extended and described in this book. DAIS (see Section 6.1) is a plan that BDIx agents execute (i.e., in the event of a D2D device entering the network) in order to select the transmission mode that the D2D device will operate. This is achieved in a distributed artificial intelligence manner and using only local network knowledge (i.e., the weighted data rate (WDR) of the D2D path, and the D2D device coordinates in proximity). Additionally, a centralized algorithmic maximization approach, called Sum Rate (SR), is proposed (in Section 6.2.1), extended to be distributed and investigated as DSR (shown in Section 6.2.2). With DSR, transmission mode selection was achieved as distributed by using global network knowledge (i.e., coordinates, data rates, transmission modes and links of all devices under the BS) and by focusing on maximizing the aggregated data rate of all the links established in the Network (we refer to this as Sum Rate). A comparative evaluation, together with other competing approaches is also offered in the book for a static environment.

Finally, in order to demonstrate the potentials of the DAI framework in a dynamic environment, we extend the enhanced DAIS approach targeting the creation of stable and efficient clusters and good backhauling links towards the gateway, considering dynamic network conditions (i.e., incorporating mobility, etc.) causing changes in the D2D network topology through subsequent TS of execution. The enhancements also highlight the extendability of DAI framework to handle other situations. To achieve this, the algorithm of enhanced DAIS plan (shown in Section 6.1.5) is extended with the speed (called MAXSpeedToFormBackhauling threshold), which restricts a device to share its link with other devices, either for cluster formation or relay traffic, according to its speed. The difficulty there is that in each time step of execution, the newly selected transmission node can affect existing clusters, as well as the formation of new clusters and backhauling links that could result in disconnected/disjointed clusters. However, these clusters and paths should not be affected, even if the UE moves away from the cluster head (CH). Moreover, we have introduced speed threshold as an extension of the enhanced DSR approach (shown in Section 6.2.4), to make it competitive, distributed, and align with DAIS in a dynamic environment. Similarly, we enhanced the SHRA approach (introduced in Ref. [26]) in order to support multiple connections at D2D-Relays and allow cluster formation. By considering mobility, these improvements are implemented within the approaches mentioned above, providing enhanced performance in terms of SE and PC and reduced computation time. A comparative evaluation, together with other competing approaches, is also offered in the book for a dynamic environment.

1.3 OVERVIEW

The rest of this book is structured as follows. Chapter 2 accommodates the background information of DAI framework and the BDI agents. Additionally, it provides a literature review of intelligent approaches in D2D and presents open issues in D2D for 5G/6G communication networks. Chapter 3 describes how the DAI framework can be realized with BDIx agents and Chapter 4 presents implementation specifics of the DAI framework for D2D communications. To exemplify the generality of the DAI framework, Chapter 5 presents several illustrative plans related to D2D challenges. Chapter 6 provides a detailed illustrative example of how the DAI framework along with other competitive techniques can be realized for D2D mode selection (frequency & transmission). Chapter 7 provides performance evaluations of the DAI framework implementation for D2D mode selection, firstly in a static and then in a dynamic environment, including a comparative evaluation with competing approaches in each environment. Chapter 8 contains work that is still in progress, conclusions, and future work. A supportive appendix is also provided in this book. The colorful figures that are related to this book are shared to the following URL: https://github.com/iacovosi/DAIBOOK.git.

REFERENCES

1. I. F. Akyildiz, A. Kak, and S. Nie, "6G and beyond: The future of wireless communications systems," *IEEE Access*, vol. 8, pp. 133995–134030, 2020.
2. I. F. Akyildiz, S. Nie, S. C. Lin, and M. Chandrasekaran, "5G roadmap: 10 key enabling technologies," *Computer Networks*, vol. 106, pp. 17–48, 2016.
3. "5G Applications and Use Cases — Digi International." [Online]. Available at: https://www.digi.com/blog/post/5g-applications-and-use-cases, Accessed on: 2021-07-24.
4. K. B. Letaief, W. Chen, Y. Shi, J. Zhang, and Y.-J. A. Zhang, "The roadmap to 6G: AI empowered wireless networks," *IEEE Communications Magazine*, vol. 57, no. 8, pp. 84–90, 2019.
5. K. David and H. Berndt, "6G vision and requirements: Is there any need for beyond 5G?," *IEEE Vehicular Technology Magazine*, vol. 13, no. 3, pp. 72–80, 2018.
6. M. Giordani, M. Polese, M. Mezzavilla, S. Rangan, and M. Zorzi, "Toward 6G networks: Use cases and technologies," *IEEE Communications Magazine*, vol. 58, no. 3, pp. 55–61, 2020.
7. E. C. Strinati, S. Barbarossa, J. González-Jiménez, D. Kténas, N. Cassiau, and C. Dehos, "6G: The next frontier," *ArXiv*, vol. abs/1901.03239, 2019.
8. R.-A. Stoica and G. Abreu, "6G: The wireless communications network for collaborative and AI applications," *ArXiv*, vol. abs/1904.03413, 2019.
9. R.-L. Aguiar, "White paper for research beyond 5G," Technical Report, October 2015.
10. S. Yrjola, "Decentralized 6G business models," in *6G Wireless Summit*, pp. 1–2, 04 2019. Finland, 2019.
11. C. Liaskos, A. Tsioliaridou, S. Nie, A. Pitsillides, S. Ioannidis, and I. Akyildiz, "An interpretable neural network for configuring programmable wireless environments," *2019 IEEE 20th International Workshop on Signal Processing Advances in Wireless Communications (SPAWC)*, Cannes, pp. 1–5, Jul. 2019.
12. ITU-T, "ITU-T Y.3170-series: Machine learning in future networks including IMT-2020: Use cases," Technical report, ITU, 2019.
13. ITU-T, "Y.3173 : Framework for evaluating intelligence levels of future networks including IMT-2020," Technical report, ITU, 2020.
14. W. Saad, M. Bennis, and M. Chen, "A vision of 6G wireless systems: Applications, trends, technologies, and open research problems," *IEEE Network*, vol. 34, no. 3, pp. 134–142, 2020.
15. E. Yaacoub and M.-S. Alouini, "A key 6G challenge and opportunity: Connecting the remaining 4 billions: A survey on rural connectivity," *ArXiv*, vol. abs/1906.11541, 2019.
16. R. Li, Z. Zhao, X. Xu, F. Ni, and H. Zhang, "The collective advantage for advancing communications and intelligence," *IEEE Wireless Communications*, vol. 27, no. 4, pp. 96–102, 2020.
17. A. M. Schweidtmann and A. Mitsos, "Deterministic global optimization with artificial neural networks embedded," *Journal of Optimization Theory and Applications*, vol. 180, no. 3, pp. 925–948, 2019.
18. M. Bratman, *Intention, Plans, and Practical Reason.* Cambridge, MA: Harvard University Press, 1987.
19. A. S. Rao and M. P. Georgeff, "BDI agents: From theory to practice.," *ICMAS*, vol. 95, pp. 312–319, 1995.

20. M. Georgeff, B. Pell, M. Pollack, M. Tambe, and M. Wooldridge, "The belief-desire-intention model of agency," in *Intelligent Agents V: Agents Theories, Architectures, and Languages* (J. P. Müller, A. S. Rao, and M. P. Singh, eds.). Berlin, Heidelberg: Springer, pp. 1–10, 1999.
21. A. S. Rao and M. Georgeff, "BDI agents: From theory to practice," in *Proceedings of the 1st International Conference on Multi-Agent Systems (ICMAS-95)*, San Francisco, CA, 1995.
22. J. Lu, L. Feng, J. Yang, M. M. Hassan, A. Alelaiwi, and I. Humar, "Artificial agent: The fusion of artificial intelligence and a mobile agent for energy-efficient traffic control in wireless sensor networks," *Future Generation Computer Systems*, vol. 95, pp. 45–51, 2019.
23. G. Fodor, E. Dahlman, G. Mildh, S. Parkvall, N. Reider, G. Miklós, and Z. Turányi, "Design aspects of network assisted device-to-device communications," *IEEE Communications Magazine*, vol. 50, pp. 170–177, 2012.
24. P. Gandotra and R. Jha, "Device-to-device communication in cellular networks: A survey," *Journal of Network and Computer Applications*, vol. 71, no. 4, pp. 1801–1819, 2016.
25. M. Ahmad, M. Azam, M. Naeem, M. Iqbal, A. Anpalagan, and M. Haneef, "Resource management in D2D communication: An optimization perspective," *Journal of Network and Computer Applications*, vol. 93, pp. 51–75, 2017.
26. U. N. Kar and D. K. Sanyal, "Experimental analysis of device-to-device communication," *2019 12th International Conference on Contemporary Computing, IC3 2019*, pp. 1–6, Noida, 2019.

Notes

[1]Note that there are a lot of predefined well structure languages for BDI agents communication.
[2]Dynamic means to react fast in a change of a situation.
[3]Flexible means adapt to possible future changes.
[4]Autonomous means having the freedom to act independently, do whatever needed in order to solve a problem.
[5]The aggregated total data rate of all the links established in the network divided by the available bandwidth of the network.

2 Background, Literature Review and Related Work

The objective of this chapter is fivefold. The first objective of this chapter is to introduce the DAI concept, including a discussion on BDI agents, which form the core of this work. The second objective of this chapter is to perform a literature review of intelligent approaches in D2D, considering those classified under the artificial intelligence (AI), machine learning (ML), and data mining (DM) fields. The third objective of this chapter is to provide a survey of open issues in D2D communications. The fourth objective of this chapter is to show the need for AI at 5G/6G and beyond. The last and fifth objective is to provide the related work that is associated with BDI agents in telecommunications, D2D frameworks, transmission mode selection, and dynamic transmission mode selection.

2.1 DAI FRAMEWORK AND BDI AGENTS

This section provides background knowledge regarding the concepts of the DAI framework and BDI agents, as reviewed from the open literature. In addition, the important architectural characteristics of the agents are provided. Finally, a description on how BDI agents can form multi-agent system is provided.

2.1.1 DISTRIBUTED ARTIFICIAL INTELLIGENCE CONCEPT

Distributed artificial intelligence (DAI) is an area of study under Al concerned with coordinated, concurrent action, and problem-solving in a distributed manner. DAI has a class of technologies and methods that span from Q-Learning to multi-agent technologies targeting the implementation of distributed approaches for a specific problem. DAI as a concept is based on intelligent agents that manage their knowledge, abilities, capabilities, and intends/plans in order to perform actions with the objective to solve problem(s) by collaboration or as individual entity for problem solving [1–4].

2.1.1.1 Areas of Distributed Artificial Intelligence

The DAI can be separated into four areas of research: Distributed Problem Solving (DPS), Parallel AI (PAI), Swarm Intelligence, and Multi-Agent Systems (MAS) [5, 6].

2.1.1.1.1 Distributed Problem Solving

The DPS investigates how a problem can be divided among several modules/nodes/agents that cooperate at the level of dividing and sharing knowledge about

DOI: 10.1201/9781003469209-2

the problem and the developing solution [5, 7]. The DPS is usually used either in constraint-satisfaction problems (DCSPs) or in distributed constraint-optimization problems (DCOPs). For each case of problems, multiple algorithms have been designed [8]. The general approach is to reduce the more significant problem into interdependent sub-tasks (spatial, temporal, or functional). The partial solutions are then integrated and joined and fit into an overall solution [9].

In DPS [6], collaboration is essential given that no individual agent has sufficient information, knowledge, and capabilities to resolve the complete problem. The designer, a researcher of DPS, has to correctly allocate the information and capabilities in such a way that agents supplement rather than conflict with each other. Typical application areas are the following (among many): (i) distributed planning and control; (ii) interpretation; (iii) cooperating expert systems; (iv) cognitive models of cooperation; and (v) human cooperation backed by digital tools.

Note that requiring cooperation increases the complexity of the system exponentially. However, according to Ref. [9]: (i) it is cheap to use, from a hardware perspective, because it allows the interconnection of multiple devices, rather than having a single centralized (equivalent in power) processor; (ii) many AI applications are distributed by nature and design; (iii) the modularization of the problem into subproblems is providing the ability to check, debug, and maintain the modules; (iv) having DPS accelerates the incorporation of AI into human society because collaboration is the evolutionary mechanism.

In order for the DPS to achieve their target and reach a solution, they need [6]: (i) to use designing incentives for the agents to work together (coherence); (ii) formulate the agents to learn how to work together (competence) via AI/ML or by standard plans.

2.1.1.1.2 Parallel AI

The parallel Al (PAI) [5] investigates how to develop parallel computer architectures, languages, and algorithms for Al. The approaches that are under the class of PAI are targeting the solution of performance problems of Al systems and do not investigate the conceptual advances in understanding the nature of reasoning and intelligent behaviour among multiple agents.

This book focuses on approaches to the problems of distributing and coordinating knowledge and action in distributed problem-solving and multi-agent systems. Note that developments in concurrent languages and architectures (PAI) have direct impact on all other areas of DAI [5].

2.1.1.1.3 Swarm Intelligence

Swarm Intelligence (introduced in Ref. [10]) investigates how the artificial natural systems made by multiple agents coordinate using decentralized control and self-organization by observing natural swarm of agents (i.e., ants). A typical swarm system has some properties: (i) it has homogeneous agents (either identical or belonging to other typologies); (ii) agents interact with each other corresponding to basic rules that only develop local information exchanged directly with another agent or via the environment (stigmergy); and (iii) the agents in the group achieve self-organization and results emerge from the overall behaviour of the system [6].

2.1.1.1.4 Multi-Agent Systems

Multi-agent systems (MAS) investigate how the intelligent coordinating behaviour among a collection of autonomous intelligent agents can coordinate their knowledge, goals, skills, and plans jointly to take action or solve problems. The agents forming a multi-agent system may be working towards a single global goal or they may be working toward separate individual goals, which make them have interaction among them. Also, agents in a multi-agent system can share knowledge about problems and solutions through collaboration [5, 6]. The agents must have reasoning about the processes of coordination among the agents. In multi-agent systems, the task of coordination can be quite difficult. Additionally, there are approaches like open systems where there is no global control, globally consistent knowledge, globally shared goals or success criteria, and global representation of a system [5]. Furthermore, in MAS, distributed autonomous agents interact with each other based on pre-determined rules/constraints and, consequently, a collective behaviour that is achieving the target solution with the use of interactions. The interactions are between the agents with other agents and between agents and the environment itself. Additionally, with the use of reinforcement learning and ML (learning basically) in the agent, the actions reward function can be maximized [6].

2.1.1.2 DAI Characteristics and Requirements

In this section, we provide the DAI characteristics and requirements [6]. The DAI can principally be used for learning, reasoning, and planning on any problem. For DAI autonomous learning, agents reach conclusions or a semi-equilibrium through interaction and synchronous or asynchronous communication and can decide with a reduced amount of data. Thus, the DAI can be defined by three main characteristics [11]: (i) It is a distribution method for the allocation of tasks between agents; (ii) it is a method of distribution of powers; and (iii) it is a method of communication of the agents.

There are specific minimum requirements by an approach to be considered distributed AI [12]. These requirements are:

1. The agents' granularity can be either acting at a task-level problem decomposition (coarse-grained) or a statement-level decomposition (fine-grained)
2. The agent's knowledge could be either redundant or specialized (heterogeneous).
3. There are several ways of distributing the control in the system (e.g., benevolent, competitive, team, hierarchical, static, shifting roles)
4. There exist different ways of communicating (e.g., blackboard model, message-model) that can be either at low or high-level content.

Essential concepts in the design of the DAI approaches are the achievement of distributed (a centralized process of task distribution) or decentralized system (allocation of tasks in a decentralized manner).

As shown in Ref. [13], in order to build a DAI, you need to have the following building blocks in the design process: (i) design the agent architecture in terms of heterogeneity, reactive and deliberate features; and (ii) design the overall distributed, autonomous system properties such as the communication channel (i.e. message-model) that the agent will use, the protocol (i.e. FIPA-ACL) and how much human will be involved in the decisions of the agent (i.e. monitoring QoE). Note that all these divisions designed for the (running) system ask the designers of the agents to make several initial checking calls using Application Programming Interface (API) e.g., Representational State Transfer (REST) targeting the coherence of agents, checking that there is a fixed protocol/language selected in order to achieve communication and interaction and finally to check that agents decision results are synthesizable and actionable.

Another design approach shown in Ref. [14] related to the DAI context is the following: The designer should analyze a system where several branches work together to achieve a common goal (following the DPS approach) or design multiple independent agents and look for an emerging solution from their interactions (following the MAS approach).

2.1.1.3 Basic Problems of Distributed AI

The basic problems of Distributed AI are the following [5]: (i) formulate, describe, decompose, and allocate problems and synthesize results among a group of intelligent agents; (ii) enable agents to communicate and interact (i.e., communication languages, protocols); (iii) assure that agents act coherently; (iv) enable individual agents to represent and reason about the actions, plans, and knowledge of other agents in order to coordinate with them); (v) recognize and reconcile disparate viewpoints and conflicting intentions among a collection of agents trying to coordinate.

2.1.1.4 DAI Control

DAI control [4, 15, 16] is a category of distributed control scheme which solves complex learning, planning, and decision-making problems in a distributed manner. With the Distributed Control described above, this DAI scheme supports perfectly parallel workload.[1]

2.1.2 BELIEF-DESIRE-INTENTION INTELLIGENT AGENTS

Intelligent agents are autonomous units that observe an environment using sensors and act upon it using actuators, coordinating their activity in the direction of achieving goals (i.e. they are "rational", as defined in economics). Agent theory is concerned with the use of mathematical formalisms for representing reasoning and the properties of agents. Software agents are characterized as computer software that display flexible autonomous behavior, which infers that these systems are capable of independent, autonomous action in order to satisfy their design objectives. Agents are utilized in a lot of applications. For instance, autonomous programs used for operator assistance or data mining (in some cases referred as bots) are also called "intelligent agents". The Belief-Desire-Intention (BDI) Agents[21], which are also called as "intelligent agents", are a category of agents with some extra functionality [22]. The features forming the extra functionality of these agents are their Beliefs, Desires, Goals, Intentions, and Behaviour. [21]. More specifically:

- The Beliefs represent a list of quantifiable and qualitative parameters, which reflect the agent's understanding (perception) of the surrounding environment. The values of these parameters are measured by the agent by considering information related to the surrounding environment. Beliefs can also include inference rules allowing advance chaining to guide toward new Beliefs and ML structures (e.g., fuzzy logic).
- The Desires correspond to the motivational state of the agent. Specifically, the Desires represent a list of objectives that the agent would like to fulfill. Each objective (referred to as Desire in the rest of the book) is associated with a specific plan, providing explicit instructions, that the agent should follow towards its realization.
- A Goal is a desire that has been adopted for active pursuit by the agent.
- The Intentions represent a list of objectives, selected (as a subset of the Desires) by the agent to perform. Specifically, the Intentions represent a list of Desires that are currently under the pursuit of the agent, either these are currently under execution or standby, following their associated plans.
- The BDI Behaviour is characterized by its Perception and the Plans included in its Planning Library and associated with its Desires. More specifically:
    - A Perception is a function executed in a case of events and sensor values change. With this function, the agent can update its Beliefs and convert Desires to Intentions according to changes identified in the environment.
    - A Plan is an algorithm that the agent must follow in the case of an event or a change of Belief to Intention.

A BDI agent decides its actions/plans, in an autonomous manner, based on its Beliefs, goals, events, and realized intentions from desires. Additionally, it is capable of interacting and cooperating with other agents based on two axes: (i) personal interest of the agent based on the Desires; (ii) the interest of the group that the agent is part of Ref. [21]. In this manner, a multi-agent system creating a collaborative environment is formed. However, two important issues to consider in multi-agent systems are the following [23]:

- Mechanisms are needed to allow agents to synchronize and coordinate their activities at runtime.
- In multiagent systems the agents are primarily concerned with their own welfare and Intentions.

Furthermore, BDI agents can communicate among them and exchange information to execute specific actions (e.g., change Intentions) or learn from other BDI agents or even instruct other agents to do a specific task.

2.1.3 IMPORTANT ARCHITECTURAL CHARACTERISTICS OF THE AGENTS

In this section, we highlight the architectural characteristics of an agent [24–26] that are considered valuable in 5G/6G Communications (i.e., persistence, priority, flexibility, responsiveness, reactivity). These characteristics are also implemented in the proposed DAI framework.

Two major characteristics of agents are *persistence* and *priority*. More precisely, the agents can have property values for persistence coefficients and priority values in their architecture. The target is, with the use of persistence, to set the level of independence to the evolutionary environment[2] with the use of a utility function. Specifically, agents that have high persistence, persist on their selected Intentions and execute their plans independently of the environment evolution and sensor input changes that affect the Beliefs. On the other hand, agents with lower persistence are adaptable, reactive, and responsive to environmental change. However, this may lead to problematic and computationally expensive behaviours. The priority characteristic of the agent is also important. Through priority values, an agent can determine the correct intention to be used from a corresponding Desire in case of a Belief change or the raise of an event.

A third important characteristic of the agent is *flexibility*. This is related to the ability of the agent to easily define and adapt its Beliefs, Desires, and Intentions (along with other agent's parameters, like Plans and Priorities), in real time. For example, an agent designer with the use of a modeler[3] can define an agent with just one Belief and some Desires and Plans that can tackle only a single problem or it can create a complex Agent that can tackle a huge problem such as the coordination of a dancing robot. Therefore, the architecture of an agent can be simple with reduced Beliefs, Desires and Plans, or complex with the use of full range of BDI components.

Another characteristic of an agent is its *responsiveness*. More specifically, the selection of the Desires that will become Intentions, is not predefined but based on the agent's behavior and responsiveness to events raised, sensor measured values, and changes in its Beliefs.

The final but equally important characteristic of an agent is *reactivity*. A reactive agent can define a cognitive model and through this model specify its target challenges along with the plans that will achieve their implementation (in the same way as in a finite state machine).

2.1.4 USE OF BDI AGENTS TO FORM MULTI-AGENT SYSTEM

BDI agents can cooperate and form a multi-agent system. Multi-agent systems are systems composed of multiple interacting computing elements capable of autonomously deciding what actions they require to perform in order to satisfy their design objectives. In multi-agent systems, the entities interact with other agents, not only by exchanging information, but also by appealing in analogues of the type of social activity that people engage in every day, like cooperation, coordination, and negotiation [27]. In multi-agent systems, there are two important issues to consider: (i) Because agents are anticipated to be autonomous it is usually expected that the synchronization and coordination structures in a multi-agent system are not hard-wired at design time, as they normally are in standard concurrent/distributed systems. In this manner, mechanisms are needed in order to allow agents to synchronize and coordinate their activities at runtime; and (ii) The encounters that occur between computing elements in a multi-agent system are financial encounters, in the sense that they are encounters between self-interested entities. In a classic distributed/concurrent system, all the computing elements are supposed (implicitly) to share a common goal (of making the overall system function correctly). In multi-agent systems, it is assumed instead that agents are primarily concerned with their own welfare (although of course, they will be acting on behalf of some user/owner) [27]. One way for such multi-agent BDI systems to communicate is by using a well-formatted standard language. FIPA (The Foundation for Intelligent Physical Agents) is an organization that defines standards for heterogeneous and interacting agents and agent-based systems. FIPA proposes the FIPA ACL, a well-defined communication language (like natural language) in order for an agent to propose something to other agents for adoption and execution, or not (if the other agents reject the proposal) [28]. Another well-formatted standard language, used widely for agents, is AngelSpeak [29].

In addition, we can say the BDI agents have foundations in the Algorithmic, Game-Theoretic, and Logical theories [27]. All the features discussed above make, in our opinion, BDI agents suitable for solving the challenges of 5G/6G communication.

2.2 DEVICE-TO-DEVICE COMMUNICATION

In this book we consider a Device-to-Device (D2D) setup in a 5G and beyond communication network to illustrate the realization of the DAI framework and exemplify its properties. In this setup, each D2D device, by controlling its cellular (i.e., LTE, 5G) and Wi-Fi interfaces, aims to achieve D2D communication. D2D can operate both in the licensed (inband D2D) and unlicensed (outband D2D) spectrum and is generally transparent to the cellular network. This is so because it allows proximate devices (UEs) to bypass the Base Station (BS) and establish direct links between them, to either share their connection and act as relay stations or directly communicate and exchange information (see Figure 2.1). As D2D allows direct communication between two devices, it promises improvements in energy efficiency, spectral efficiency, overall system capacity, higher data rates, efficient offloading and load balancing, license-exempt band, controllable interference in the licensed spectrum, and due to the allowed communication distance of 5–50 m, and low power consumption [30–34].

The target is to tackle the D2D challenges by focusing on the local environment of D2D communication, rather than the global environment. Additionally, relying only

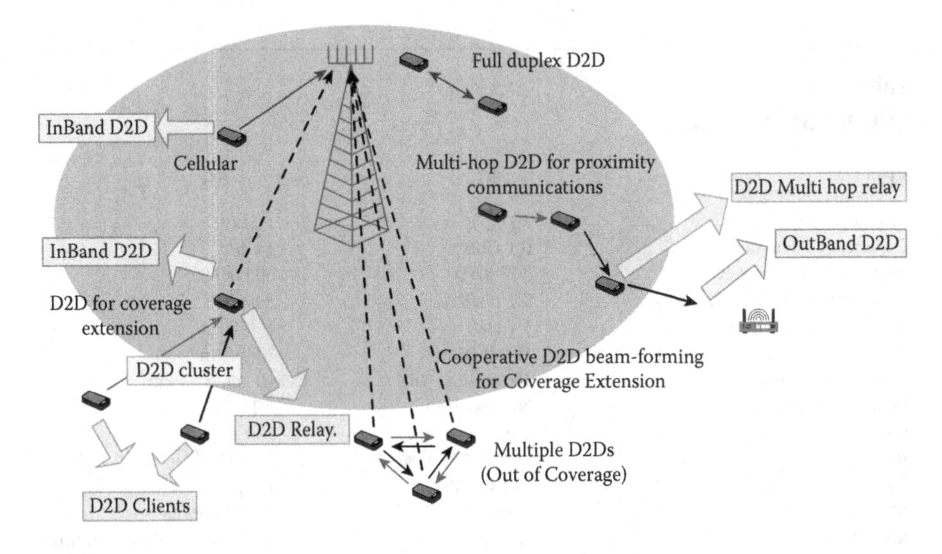

Figure 2.1 Device-to-device communication.

on the local environment, it is expected to result in reduced signalling overhead and much faster control decision.

This section provides the following: (i) Primary Key Performance Indicators (KPIs) for evaluation of the 5G/6G D2D approach; (ii) the D2D technical challenges that an approach must solve to achieve 5G/6G; (iii) the methodology used in the literature review of intelligent approaches tackling D2D communication; (iv) identified intelligent approaches categorized per group of technology used; (v) general observation on intelligent approaches; (vi) taxonomy of groups based on spectrum utilization, transmission mode and control; (vii) overview of the literature review; and (viii) concluding remarks and open issues identified.

2.2.1 KPIS: KEY PERFORMANCE INDICATORS

The demanding requirements of D2D communication must meet very stringent performance criteria, as given by the standards and the research communities. Table 2.1, shows the Key Performance Indicators (KPIs) for 5G and 6G.

2.2.2 D2D TECHNICAL CHALLENGES

In order for D2D to mature and shape the D2D communication for the 5G and beyond wireless communication network, a number of technical challenges and issues must be resolved. These include aspects related to Device Discovery, Mode Selection, Interference Management, Power Control, Security of D2D communication, Radio Resource Allocation, Cell Densification and Offloading, QoS / Path Selection (Routing), D2D using mmWave communication, and Handover of D2D devices

Table 2.1
KPIs for 5G and 6G

KPI	5G	6G
Peak data rate	20 Gb/s	1 Tb/s
Experienced data rate	0.1 Gb/s	1 Gb/s
Peak spectral ef?ciency	30 b/s/Hz	60 b/s/Hz
Experienced spectral ef?ciency	0.3 b/s/Hz	3 b/s/Hz
Maximum bandwidth	1 GHz	100 GHz
Area traf?c capacity	10 Mb/s/m^2	1 Gb/s/m^2
Connection density	10^6 devices/km^2	10^7 devices/km^2
Energy ef?ciency	Not speci?ed	1 Tb/J
Latency	1 ms	100 μs
Reliability	$(1-10^{-5})$ in %	$(1-10^{-9})$ in %
Jitter	Not speci?ed	1 μs
Mobility	500 km/hour	1000 km/hour
Reconfiguration	Not specified	Re-configurable in real time

Source: Adapted from Ref. [35].

[36,37]. How each one of these D2D technical challenges and issues is addressed by the intelligent D2D approaches is elaborated below.

2.2.2.1 Device Discovery

In order for two devices (i.e., UE/UEs) to directly communicate with one another, they must first perform the device discovery process to identify that they are close to each other and within range for D2D communication. The device discovery (DD) includes the sending of a discovery signal aiming to identify the presence of possible devices in proximity [31]. When two devices are found in range for D2D communication, these are considered as D2D candidates. Then, a series of messages about link quality is exchanged between devices and the BS, or directly between the devices. This information is considered important because it serves as the basic input to the Mode Selection criterion (see below) that should be satisfied in order for D2D candidates to be able to directly communicate [31,37].

2.2.2.2 Mode Selection

When a pair of D2D candidates identifies each other for possible future communication, Mode Selection (MS) is performed. Mode selection implies that a decision is made whether the D2D candidates will communicate directly or via the network as conventional cellular network [37,38]. Note that the Communication Mode Selection should be carefully chosen as it has a direct impact on the interference in the network.

Mode selection can be separated into two parts: (i) Frequency Mode Selection/Spectrum Utilization and (ii) Transmission Mode Selection. Below, the different modes in which a D2D device can operate, are outlined.

2.2.2.2.1 *Types of Frequency Mode/Spectrum Utilization in D2D Communication*

The types of frequency modes (spectrum utilization) [37] that can be used for the establishment of D2D Communication links, are categorized as follows:

- With Inband Overlay, a rigid fraction of the licensed spectrum is reserved for D2D UEs. This is important as one band should be kept for emergency use when a UE has to communicate due to an incident (e.g. car accident, ambulance) with special rights.
- With Inband Underlay, D2D communication takes place over the same licensed spectrum intended for legacy cellular simultaneously. This is important since the D2D devices and other UEs can reuse the same bands, thus enhancing overall spectral efficiency and improving capacity.
- With Inband Cellular, a D2D device can use in some cases its cellular resources to establish a D2D Communication link without interfering with the BS (i.e., D2DMHR).
- With Outband Controlled, D2D UEs can exploit unlicensed spectrum to establish a D2D cluster between other D2D devices and the BS. This is important since in this case a shared link between a number of D2D devices and the BS can be established, the communication of the clustered D2D devices can be local and bypass the BS. Consequently, there is a significant saving of resources.
- With Outband Autonomous, D2D UEs exploit unlicensed spectrum to communicate and they utilize other non-cellular access points (e.g., Wi-Fi) than the BS, increasing thus, the total sum rate in the network.

2.2.2.2.2 *Types of Transmission Modes in D2D Communication*

Different transmission modes exist for D2D communication, based on how D2D devices interact with the BS or between each other (see Figure 2.2). The Transmission modes [37] that D2D devices can operate are explained below:

- D2D Direct (D2DD): Two D2D devices connect and communicate directly with each other by utilizing licensed or unlicensed spectrum.
- D2D Backhauling: Achieved by D2D Single-hop or Multi-hop Relaying.
    - D2D Single-hop Relaying/D2D Relay (D2DSHR): One of the D2D devices is connected to a BS or Access Point and provides access, by sharing its bandwidth, to another D2D device/devices [39].
    - D2D Multi-hop Relay (D2DMHR): With this mode the single-hop relaying mode is extended by empowering the connection of more D2D devices as a bridge in path to achieve both backhauling and/or D2D transmissions [40].

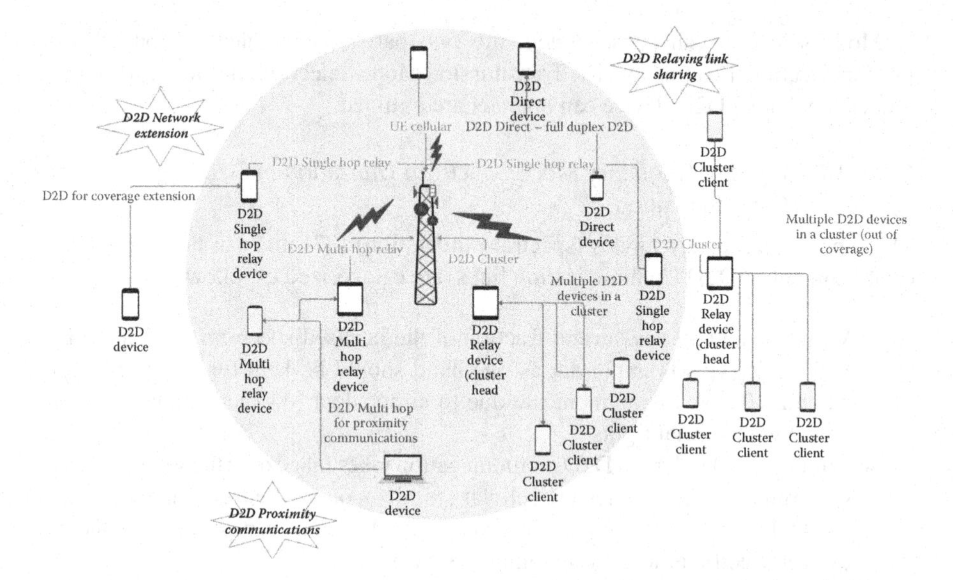

Figure 2.2 Types of transmission modes in D2D communication.

- D2D Cluster: A D2D device operating as a D2DSHR acts as a Cluster Head (CH) or Group Owner (in Wi-Fi Direct), sharing its bandwidth with one or more D2D devices [41, 42]. The Cluster Head (CH) acts as an intermediate router to the network through an access point or BS. Clustering is appropriate in high user densities.
- D2D Client (D2DC): The D2D devices connected to a D2D Cluster are called D2D Clients or Group Members (in Wi-Fi Direct) [41,42].

2.2.2.3 Interference Management

One of the most important challenges of D2D communication in cellular networks is Interference Management (IM). As indicated above, the communication Mode Selection has a direct impact on the interference in the network, especially when spectral efficiency is favored by the Network Operators. For example, when the Reuse/Underlay resource-sharing mode is selected, spectral efficiency can be achieved, however, since many D2D and cellular users will use the same portion of spectrum, the interference problem will be increased. This additionally generated interference, if not well controlled, may negate any potential benefits of D2D communication, since the overall cellular capacity and efficiency will be degraded. As an example, Interference Management in D2D Communication can be achieved by using diverse interference mitigation techniques [37].

2.2.2.4 Power Control

Although high transmission power can provide wider coverage and better signal quality during D2D communication, it can, at the same time, drain the battery of D2D UEs and cause interference to the network. In addition, Power Control (named P-C) is an important factor when other users exploit and drain one's battery when acting as a D2DSHR. Thus proper power control during D2D communication is vital for controlling the transmit power levels of D2D UEs, so as to deal with the interference generated by the D2D UEs, to improve spectral efficiency, system capacity, coverage, and reduce energy consumption [43–45]. In addition, battery power control is an important factor for user experience and the continuity of the formed D2D communications network (e.g. as a D2DSHR Node).

2.2.2.5 Security of D2D Communication

In D2D communications the routing of users' data is made through other users' devices must also consider Security (S). This makes the D2D Communication network vulnerable to many security risks and malicious attacks (see below) that could breach the data privacy and confidentiality. In D2D communication, we can have many forms of malicious attacks like: (i) eavesdropping; (ii) man-in-the-middle; (iii) free riding; (iv) denial of service; (v) node impersonation; (vi) malware attacks; (vii) Internet Protocol (IP)/bandwidth spoofing; (viii) inference attack; (ix) trust forging; and (x) location spoofing. Thus, providing efficient security (e.g., improved authentication and key agreement mechanisms) is a major issue in order to secure D2D communication in cellular networks. It is worth highlighting that interference exploitation can be used as an aid to provide secret communication in D2D communication, as proposed in Refs. [46–48].

2.2.2.6 Radio Resource Allocation

Another major issue that D2D communication needs to tackle is Radio Resource Allocation (RRA) [37]. Radio Resource Allocation mainly addresses the issues of how to assign the frequency resources to a group of D2D pairs, or all the D2D pairs. The aim is to achieve an optimal use of the radio resources focusing also on the interference control and management between D2D and Cellular links and the efficient reuse of the radio resources whenever the interference is small. However, to realize the full potential of D2D communications, the Radio Resource Allocation should be done jointly with the Mode Selection and Power Control. The purpose is to utilize the limited radio-frequency spectrum resources and radio network infrastructure as efficiently as possible. Radio Resource Allocation concerns multi-user and multi-cell network capacity issues, rather than the point-to-point channel capacity [37].

2.2.2.7 Cell Densification and Offloading

Providing high system capacity and high per-user data rates–basic requirements for 5G/6G networks–will require a densification of the radio access network or the deployment of additional network nodes. In general, the idea of network densification [49] and offloading (CDO) for performance enhancement directs the deployment of small coverage cells (e.g., Picocells and Femtocells) within a close distance to the terminal/devices, leading to additional favorable channel conditions between transmitters and receivers. Hence demands in transmission power are reduced resulting in less interference towards different co-existing network parts and thus further improvements in achievable data rates [37]. Moreover, with the massive growth in the use of smart phones and tablets, the core and access networks tend to overload. Thus, with these scenarios, offloading of cellular data is an important concern of operators, so as to free up the loaded path by providing alternate paths to the traffic. An efficient means for offloading the traffic can be provided by small cells (e.g., Picocells and Femtocells [43]) as there is less competition among the users for resources, yielding a substantial increase in spectrum efficiency. However, another offloading technique which also results in an enhancement in network capacity is D2D communication. D2D offloading avoids radio congestion as well, apart from offloading the core network. Note, however, that D2D communication mainly focuses on offloading proximity services while small cells focuses on offloading hot-spot traffic. In addition, to offloading distributed ultra-dense networks is a major part of 5G and it is used in order to tackle CDO in 5G [50]. Therefore, D2D communication should consider implementing ultra-dense networks by utilizing the D2D Cluster transmission mode. So with D2D, a cluster of D2D UEs is formed under a D2DSHR Cluster Head. The cell densification can occur where the resources (i.e., bandwidth) that are shared are limited, and in the mobile network, always the resources are limited. Therefore, the D2D Cluster/Cell densification must be implemented for the D2D communication. In order to maximize the densification in D2D communication network the maximum supported number of D2D UEs must be allocated and allowed access under one cluster (restricted by the protocol Wi-Fi Direct is 250). However, this may create at the same time, many requests trying to access a cluster; therefore, an offloading mechanism must exist in order to redirect D2D UE devices to other D2DSHR nodes. In addition, Cluster/Cell Densification can happen when a user at a cluster overuses the bandwidth shared from the CH. In such cases, an offloading mechanism must also exist in order to send the excess of the D2D clients to another D2D cluster or the D2D UE that overuses traffic to a device that can handle the excess request.

2.2.2.8 QoS/QoE - Path Selection/Routing

During D2D Communication it is essential to ensure that the QoS and QoE (QoS'P or QoS/QoE) requirements of the communication links are satisfied. To achieve this, a major issue is the selection of the optimum routing path, otherwise excess resources/power/link usage (bandwidth) will be wasted. So optimum path selection should be considered when a solution for D2D is implemented [44, 51, 52].

2.2.2.9 D2D Using mmWave Communication

Communication using mmWave (mmW) band has recently received significant attention for 5G/6G cellular networks and D2D communication, as it operates on a much higher frequency band (30–300 GHZ); thus, allowing an enormous increase in data rates (multi-Gbps) and network capacity. However, mmWaves communication suffers in terms of high propagation loss, directivity, and sensitivity to blockage, requiring Line-of-Sight (LOS) paths in order for two devices to be able to communicate. These characteristics of mmWave communications must face several challenges in order to completely develop the promise of mmWave communications, including integrated circuits and system design, interference management, spatial reuse, anti-blockage, and dynamics control [53], or even the concept of Intelligent Programmable Wireless Environments [54].

2.2.2.10 Handover of D2D Device

In order to keep the communication between two D2D devices, when these are moving away from each other, Handover (H'D2D or HO) to alternative connections should be performed. For example, when a D2D device is moving away from the access point (e.g., a D2D Relay (D2DSHR) or a D2D Cluster Head) that is supported by, then the problem of handing the device over to the best available access point with a shared medium should be tackled [44].

2.3 SURVEY ON D2D INTELLIGENT APPROACHES

Despite the significant amount of papers and excellent reviews addressing the D2D challenges referred above, a thorough review focusing only on intelligent D2D approaches is missing and will be beneficial, especially for identifying promising solutions and open problems.

2.3.1 ADOPTED METHODOLOGY

Before describing the reviewed work, the methodology in collecting papers and handling all the information from the papers is outlined. It involves the following six steps: (i) Collection of the state-of-the-art work; (ii) grouping of related work; (iii) Analyzing and Evaluating related work; (iv) Extracting conclusions on knowledge; (v) Identifying promising solutions; (vi) Identifying open research problems and conclude on a road-map, and finally (vii) proposing an intelligent approach that will take under consideration the road-map and proposing a solution that it tackle the D2D challenges and implement D2D communication in 5G. After the identification of open research problems and conclusion of the road-map, this research focuses on utilizing the road-map guidelines proposed and conclude in a framework that can successfully implement 5G D2D communications. More specifically :

Step 1 - Collection of the State of the Art Work: As a first step, a keyword-based search for conference papers and articles was performed in well-known scientific databases (e.g., IEEE Xplore, ACM, DBLP, ScienceDirect, CiteSeerX, Wiley), and search engines. Various keywords were used such as "D2D", "Device2Device", "Device to Device", "intelligent", "device to device", "Intercell", "small cells", "hetnets", "fuzzy logic", "Q", "Q-learning", "Neural Networks","Bayesian", "Thomson sampling", "Thompson sampling and Bayesian control", "evolution algorithms", "Genetic Algorithms", "PSO", "Particle Swarm Optimization", "ACO", "Ant Colony Optimization", "artificial intelligent algorithms", "machine learning algorithms", "data mining algorithms" and combinations of them. In addition to the above selection criteria, an additional criterion is to demonstrate a convincing proof of concept by using simulations or emulations. Existing surveys on D2D [37,48,55–60] were also studied for relevant efforts and for quickly identifying the state of art. The focus was to pick only papers which followed the main concepts, design principles and challenges of the D2D. Also papers that used proprietary protocols and in which vendor lock-in was evident, were excluded from our research. This means that some popular and highly cited papers in D2D might have been excluded from our research (e.g., [61, 62]). In addition, it is worth highlighting that this research and more specifically the concluded road-map focusses on the establishment of communication of D2D devices and not in using the concept of cache and social networks in D2D. Therefore, this research does not consider Big Data cache solutions [63, 64], Q-Learning social network solutions [65] and ML similarity-based solutions [65]. Towards this end, 85 papers in total were firstly identified. An in depth search was then performed on their most relevant references, increasing the list of papers to 160. Out of these 160, 35 papers that were investigating the D2D communications in an intelligent manner, were selected and analyzed further.

Step 2 - Grouping of Related Work: In the second step, related work was categorized in groups. More precisely, eight groups were created (see Section 2.3.2), in which the papers were placed according to the intelligent approach they exploit to address a D2D challenge (i.e., fuzzy logic, Q-Learning, etc.). Note that in case a paper combined a mix of intelligent approaches in its solution this was grouped based on the main approach used. Then, all groups were further put in taxonomy of approach used: (i) Spectrum utilization (i.e., Inband or Outband) for establishing the communication link; (ii) The way Control is performed (i.e., centralized, semi-distributed, distributed) for establishing D2D communication; and (iii) the D2D Transmission mode allowed (i.e., D2D relay, D2D cluster, D2D multi hop relay) for D2D communication.

Step 3 - Analysis and Evaluation of Related Work: In this step, each paper included in each group was examined and analyzed, recording its summary, the approach used as well as the D2D Challenges and the way the D2D Challenges are addressed by each approach. Based on the analysis performed, the importance of each intelligent approach is evaluated based on: (i) its overall novelty and importance; (ii) the following attributes: the way control is performed (i.e., central, semi-distributed, distributed, Distributed Artificial Intelligence (DAI)) for establishing D2D communication; reduced complexity during establishing D2D Communication (e.g., support multiple subnets under the Cellular Network); achieving optimization and fast decisions; facilitating dynamic behavior and flexibility on changes promoting self-healing; and reduced messaging exchange; and (iii) its impact to the D2D challenges. After this step, the most significant papers per group were selected and compared, and conclusions on knowledge were extracted.

Step 4 - Extract Conclusions on Knowledge: Based on the analysis and evaluation performed on the papers in each group, conclusions were extracted. To aid this process, graphs and tables were constructed demonstrating what aspects of D2D challenges are satisfied by each intelligent approach. Also the merits and shortcomings of each approach are highlighted.

Step 5 - Identify Promising Solutions: Based on the extracted conclusions, the most promising intelligent approaches (based on our analysis and understanding) able to address the open issues and satisfy the D2D Challenges are identified.

Step 6 - Identify Open Research Problems and Propose a Road-map: In this step the identified open research problems, extracted in step 4 and 5 are combined in order to conclude in a road-map. The road-map can be used to motivate readers toward the next steps that can address the open issues and hence satisfy the D2D challenges.

Step 7 - Propose a Solution that can Tackle the D2D challenges and Implement D2D Communication with the Satisfaction of 5G Requirements: In this step, given the above analysis, we focus on the suggested implementations of the road-map, produced in step 6. To this end, to successfully implement the D2D challenges, we propose a flexible and dynamic AI/ML framework for the implementation of a distributed autonomous control environment.

2.3.2 INTELLIGENT APPROACHES FOR D2D

In this section, we investigate the adoption of intelligent approaches in D2D. An important aspect from the literature is that some approaches utilize agents (i.e., Q-Learning), some of them utilize Deep Learning and some Reinforcement Learning [66–68]. There are a lot of intelligent approaches that aim to tackle the D2D problem of communication; this research focuses on the wider field of intelligent approaches and investigates how these are used in the field of D2D Communications to address the D2D challenges. In this road-map, intelligent approaches [69] are considered all those that are used in the AI, ML, and Data Mining (DM) fields. This includes fuzzy logic [70], Q-Learning [71, 72], Neural Networks [73], Thompson sampling and Bayesian control [74–76], Evolutionary Algorithms [77–79], Genetic Algorithms [77–79], Particle Swarm Optimization [80] and Ant Colony Optimization [31]. A brief description of the logic of these intelligent approaches is provided below.

2.3.2.1 Fuzzy Logic

Fuzzy logic (FL) [44, 52, 81] is one of the fields in Artificial Intelligence (AI) which has gained importance and popularity over the last few decades. Fuzzy logic is a multivalued logic, which allows intermediate values to be defined between conventional evaluations like true/false, yes/no, high/low, etc.; is an approach to computing based on "degrees of truth" rather than the usual "true or false" (1 or 0) Boolean logic on which the modern computer is based. FL deals with reasoning that is approximate rather than fixed or exact. The base of FL is fuzzy set, which is basically a prolongation of classical set. A Fuzzy set can be best understood in the context of set membership. Basically, it allows partial membership, which means that it contains elements that have varying degrees of membership in the set. On the other hand, a Classical set contains elements that satisfy precise properties of membership. Thus, by introducing the notion of degree in the verification of a condition (thus enabling a condition to be in a state other than true or false) fuzzy logic provides a very valuable flexibility for reasoning, which makes it possible to take into account inaccuracies and uncertainties. One advantage of fuzzy logic in order to formalize human reasoning is that the rules are set in natural language, fuzzy if-then rules or, simply, fuzzy rules [82]. Fuzzy logic is essential to the development of human-like capabilities for AI, sometimes referred to as artificial general intelligence: the representation of generalized human cognitive abilities in software so that, faced with an unfamiliar task, the AI system could find a solution.

2.3.2.2 Q-Learning

Q-learning (QL) [66–68, 83–89] is a reinforcement learning technique used in ML. The goal of Q-Learning is to learn a policy, which tells an agent what action to take, and under what circumstances. It does not require a model of the environment and can handle problems with stochastic transitions and rewards, without requiring adaptations. In addition, it is a form of model-free reinforcement learning. It can also be viewed as a method of asynchronous dynamic programming (DP). It provides agents with the capability of learning to act optimally in Markovian domains by experiencing the consequences of actions, without requiring them to build maps of the domains. In Q-learning an agent tries to learn the optimal policy from its history of interaction with the environment. A history of an agent is a sequence of state-action-rewards [72, 90, 91]. Deep Learning is a specific method used to train and build Q networks. More precisely deep Q-Learning uses the power of deep learning, specifically neural networks, to predict the q-values of the different states (Further details about deep learning appear in the Neural Networks section).

2.3.2.3 Neural Networks

A Neural Network (NN) [92–98] consists on neurons that have inputs and outputs. More precisely, a neural network represents a connected graph with input neurons, output neurons, and weighted edges. The input neurons do not have any predecessor neurons and they have output neurons. In addition, the output neurons do not have any successor neuron and they have inputs. Neurons of a neural network are connected by using connections (edges), each connection transferring the output of a neuron to the input of another neuron. Each connection (edge) is assigned a weight. There are many NN approaches, a common example is the back propagation NN. In this type of NN the propagation function computes the input of a neuron from the outputs of predecessor neurons. The propagation function is leveraged during the forward propagation stage of training. The learning rule is a function that modifies the weights of the connections. This serves to produce a favored output for a given input for the neural network. The learning rule is leveraged during the backward propagation stage of training. A popular NN lately is the Deep Neural Network. It has more layers than smaller Neural Networks. A smaller Neural Network might have 1–3 layers of neurons. On the other hand, a Deep Neural Network (DNN) has more than a few layers of neurons. A DNN might have above 20 layers of neurons [95, 99, 100].

2.3.2.4 Thompson Sampling and Bayesian Control

Thompson sampling (TB) [101] is a heuristic approach that is combining probability theory and causal interventions for choosing actions that address the exploration-exploitation for solving challenging problems [74, 94]. It consists on choosing the action that maximizes the expected reward with respect to a belief that is randomly selected. In this approach, there is a player with a set of contexts, a set of actions, and rewards. In each round, the player obtains a context, plays an action, and receives a reward following a distribution that depends on the context and the issued action. The aim of the player is to play actions such as to maximize the cumulative rewards. A more general Thompson sampling that is used to arbitrary dynamical environments and causal structures is the Bayesian control rule. In this formulation, an agent is conceptualized as a mixture over a set of behaviors. As the agent interacts with its environment, it learns the causal properties and adopts the behaviour that minimizes the relative entropy to the behaviour with the best prediction of the environment's behaviour. If these behaviors have been chosen according to the maximum expected utility principle, then the asymptotic behaviour of the Bayesian control rule matches the asymptotic behaviour of the perfectly rational agent [76, 102–104].

2.3.2.5 Evolutionary Algorithms

An evolutionary algorithm (EA) [105] is a subset of evolutionary computation, a generic population-based metaheuristic optimization algorithm [106]. The EA workings are inspired by biological evolution, such as reproduction, mutation, recombination, and selection. Proposed solutions to the optimization problem are individuals in a population of solutions, and the fitness function determines the quality of the solutions. Evolution of the population then takes place after the repeated application of the above operators. Specifically, EA contains four steps: (i) initialization (initial population of solutions), (ii) selection (members of the population must now be evaluated according to a fitness function. A fitness function is a function that takes in the characteristics of a member, and outputs a numerical representation of how viable a solution is), (iii) genetic operators (crossover and mutation), and (iv) termination (repeat until termination condition satisfied: Select best fit reproduction, selection, genetic operators and Replace least-fit population). These steps each correspond, roughly, to a particular facet of natural selection, and provide easy ways to modularize implementations of this algorithm category. Simply put, in an EA, fitter members will survive and proliferate, while unfit members will die off and not contribute to the gene pool of further generations, much like in natural selection. Evolutionary algorithms often perform well approximating solutions to all types of problems, and more especially combinatorial problems, because they ideally do not make any assumption about the underlying fitness landscape. EAs are maintaining a population of potential solutions and in some way artificially 'evolving' that population over time. Some categories of EAs are: (i) Genetic Algorithms (GAs); (ii) Genetic Programming (GP), and Evolution Strategies (ES). EAs are flexible and they can address any optimization task. However, with the supported flexibility the cost for performing EA is high. So, tailoring EA's configuration and parameters, in order to reduce costs, is often a complex and time-consuming process. This tailoring process is one of the many ongoing research areas associated with EAs. In addition, EAs have computational complexity which is a prohibiting factor. The computational complexity is due to the fitness function evaluation. Fitness approximation is proposed as one of the solutions to overcome this difficulty [106, 107].

2.3.2.6 Genetic Algorithms

Genetic Algorithms (GA) [108, 109] are stochastic search-based algorithms which use the concepts of natural selection and genetics as found in nature. Note that even though GAs are a subset of Evolutionary Computation algorithms in our analysis we considered this Intelligent approach as a different group. The GAs are fast and mostly provide good results, in comparison to other algorithms, but due to their stochastic nature, the algorithm does not quarantine the quality of the result. Moreover, GAs are not suitable for simple problems which derivative information is available. Because GA are based on the process of evolution by natural selection, which has, been observed in nature, they can be used to design computer algorithms, to schedule tasks, and to solve other optimization problems. They replicate the way that life uses evolution to find solutions to real world problems. That is why GAs can solve complicated problems. The GA uses a genetic representation of the solution domain and a fitness function to evaluate the solution domain. The process steps in a GA are: (i) Initialization: In this step, the GA creates an initial population; (ii) Evaluation: In this step, each member of the population is evaluated and a 'fitness' value for each is calculated. The fitness value is calculated by how well it fits within the desired requirements; (iii) Selection: In this step the GA discards the bad designs and keeps the best individuals in the population. The aim is to constantly improve the overall fitness of the populations; (iv) Crossover: In this phase, new individuals are created by combining aspects of selected individuals. The aim is to create an even "fitter" offspring, which will inherit the best traits from each of the parents; (v) Mutation:

In this step, the GA makes very small changes at random to an individual's genome. The aim is to add a little bit randomness into the populations' genetics otherwise every combination of solutions created would be in the initial population (allows exploration); (vi) Repeat!: As the next generation has been created, the algorithm starts again from step two until a solution, which meets a predefined goodness criteria, is reached. Some limitations of GAs are that the computation of fitness value might be extensive for some problems and may not converge to the optimal solution. On the other hand, the advantages of GA are that it can run in parallel, does not require any derivative information, is fast and gives a good solution. Moreover, it always has an answer for the investigated problem [106, 110].

2.3.2.7 Particle Swarm Optimization

Particle swarm optimization (PSO) [111–118] is a population based stochastic optimization technique inspired by social behavior of bird flocking or fish schooling. PSOs have many similarities with Genetic Algorithms (GA). The system is initialized with a population of random solutions and searches for optima by updating generations. However, PSO has no evolution operators such as crossover and mutation. The potential solutions in PSO are called particles, which fly through the problem space by following the current optimum particles. These particles are moving around in the solution search-space according to a mathematical formulae over the particle's position and velocity. Each particle's movement is influenced by its local best known position solutions. Therefore, it is guided toward the best known positions in the search-space, which are updated as better positions are found by other particles. This movement makes the swarm move toward the best solutions. The way the movement is executed is the following: At each time step, each particle acceleration is weighted by a random term, with separate random numbers being generated for acceleration toward the best locations. In addition, in the PSO there are few parameters to adjust. Parameters have also been tuned for various optimization scenarios. The choice of PSO parameters can have a large impact on optimization performance [119–121]. The negative factor in this approach is the use of the random term and the separation of random numbers. This causes delays in the calculation of the final output.

2.3.2.8 Ant Colony Optimization

The ant colony optimization algorithm (ACO) [122, 123] is a probabilistic technique for solving computational problems that are reduced to finding the best path through a graph. This approach can be classed as a Computational Intelligence (CI) technique [69], whereby a colony of artificial ants cooperates to solve discrete optimization problems. The artificial ants (simulation agents) are multi-agent methods that aim to replicate the behavior of real ants. The AI ants locate optimal solutions by moving through a parameter space representing all possible solutions, mimicking real ants in the sense of laying down pheromones as they move to a target (e.g., food source), thus directing each other through the pheromone concentration to (food) resources while exploring their environment. The AI ants similarly record their positions and the quality of their solutions, so that in later simulation iterations more ants are 'attracted' and can locate better solutions. A variation CI approach is the bees algorithm, which is more analogous to the foraging patterns of the honey bee [124–127].

2.3.3 GENERAL OBSERVATIONS PER INTELLIGENT APPROACH

Before analyzing and categorizing related work into groups of intelligent approaches, the aim of this section is to summarize the realizations of each group. In the summary of each intelligent group some characteristics related to the group are investigated and analyzed. More specifically, the characteristics upon which each Intelligent Approach is analyzed are the D2D challenges addressed, features supported, type of control used, spectrum utilization and transmission modes allowed. Any Intelligent approach must be dynamic, flexible and autonomous. Moreover, the flexibility that is the ability for the approach to adapt to possible, future changes in its requirements (i.e., increase the number of D2D devices, add mmWaves communication, D2DSHR goes offline) and react fast in a change of a situation (i.e., a D2D device enters/leaves the D2D network). Also, dynamicity that is the characteristic of the approach to react to changing conditions of operation (i.e., D2D device change coordinates, increase speed, etc) and continue satisfying the D2D Challenges.[4] In addition, the autonomicity that the approach is having the freedom to act independently in order to solve a problem of each Intelligent Approach is analyzed and some general observations are provided. Note that the last five characteristics are the same that BDI agents support.

2.3.3.1 FL Group

The majority of D2D communication approaches that use fuzzy logic mainly focus on the technical challenge related to Handover of D2D device in Heterogeneous Networks, and to a lesser extent also address D2D challenges related to: Device Discovery, Interference Management, Power Control, Security, Radio Resource allocation, Cell Densification and offloading, and QoS. In a number of papers, in order to jointly satisfy the successful implementation of the approach and address some of the aforesaid D2D challenges, FL is utilized with other networking technologies (like Software Defined Network (SDN), Network Function Virtualization (NFV), L7 Switch, OpenFlow and Cramer-shoup KEM) or supplemented with other intelligent approaches (like Ant Colony Optimization as a secondary complementary technique). More specifically, the implementation of SDN and NVF in some fuzzy logic approaches were used for facilitating Device Discovery, Interference Management and Radio Resource Allocation, while the Cramer-shoup key encapsulation mechanism (KEM) technique was used for security of the D2D communication links. Cell Densification and offloading with QoS was facilitated with the use of L7 Switch, the implementation of OpenFlow protocol and the use of Ant Colony Optimization as a secondary complementary technique. Technique for Order Preference by Similarity to Ideal Solution (TOPSIS) technique and Adaptive Neuro-Fuzzy Inference System (ANFIS) architecture were used for the Handover of D2D device. General observations: Fuzzy logic approaches as such cannot be considered flexible but they are dynamic. The fuzzy logic approaches can handle dynamic situations (i.e., like a D2D UE location change) by using a dynamic rule base on the thresholds the approaches pre-defined. However, they are not flexible, as they cannot adapt fast to changes in the topology of the D2D Network as the algorithms must be rerun. More specifically, for any small change in D2D network (i.e., addition of a new Candidate UE Device to use D2D), the algorithms must rerun to recalculate the network frequencies and transmission power of the D2D devices, which takes time for the system to adjust. This may degrade the system since the execution of the algorithms requires an extensive use of Central Processing Unit (CPU), battery and network bandwidth. Also they cannot be considered as autonomous since the D2D devices cannot act independently as they have to follow the guidelines set by the estimated thresholds. Although, as shown in Refs. [44, 52], the framework of SDN and NFV enables real time management, flexibility, and automaticity (i.e., allow the D2D devices to act autonomously and create a distributed network).

2.3.3.2 Q-Learning Group

The D2D communication approaches that use Q-Learning mainly focus on the D2D challenges of QoS and Power Control, and to a lesser extent also address D2D challenges related to: Mode Selection, Interference Management, Security and Radio Resource Allocation. In some cases the Q-Learning, in order to jointly satisfy the successful implementation of the approach and address some of the aforesaid D2D challenges, is utilized with stochastic theory. More specifically, the implementation of stochastic theory in some Q-Learning approaches was used for facilitating Interference Management and Radio Resource Allocation. It is worth noting that the trends in D2D communication research appear to favour Q-Learning. General observations: Q-Learning approaches can handle dynamic situations (i.e., like a D2D UE location change) [89] by forcing a generation of a positive/negative reward in each change. However, they are not flexible, as they cannot adapt fast to changes in the topology of the D2D Network as the algorithms must be rerun (the Markov approach is a base of the Q-Learning theory). Additionally, even if Q-Learning approaches, by default, support autonomous nodes and agents, we could not identify any Q-Learning approach in the literature that was using the D2D device as autonomous or using the full potential of agents. Moreover, even though Q-Learning is using distributed control, the Q-Learning approaches found in the open literature did not implement DAI, as they depended on the BS (or in a case of distributed control by the D2D devices using BS information) to calculate in advance, some thresholds or limits (like bit rate). However, these may be high-level directivesset points for the guidance/operation of the devices. If the thresholds are not set up at any particular period they cannot operate independently because the decisions will take are critical for the whole convergence of the algorithm as they vary in the decision process. Therefore, thresholds need to setup correctly in order to do a successful run. Because the control used in the group approaches is not multilevel control where in this type of control the approach can use the thresholds to in another level of control. An advantage of Q-Learning approaches is that they can have a history on the reconciliation factor for decision making (e.g., for selecting the best solution with the lower interference). In addition, because Q-Learning depends on action award, this may restrict usage of some other intelligent and non-intelligent approaches, like Game Theory. The reason is that Game Theory needs a specific direct response of other entities, which cannot be achieved by Q-Learning. Also, although not found in the open literature, Q-Learning approaches can utilize any other intelligent approach (i.e., fuzzy logic) in its utility function and help in addressing more D2D challenges (e.g., Handover with fuzzy logic). Some of the approaches using Q-Learning are fast to conclude on the task by using state-action–reward approach. However, it is important to setup the learning rate correctly, in order not to misbehave. Because the execution of Q-Learning algorithms is time- consuming, due to trial and error, it takes time to conclude when the learning rate is not setup correctly. Therefore, the re-run execution is considered expensive as there is an extensive use of CPU, battery, and network bandwidth.

2.3.3.3 Neural Network Group

The D2D communication approaches that use Neural Networks mainly focus on the D2D challenges related to Interference Management and Power Control, and to a lesser extent also address D2D challenges related to: Radio resource allocation, mmWave, and Handover of D2D devices. Note that this group is the only one that handles the D2D technical challenge related to mmWave in D2D Communication. Recent trends in D2D communication research appear to be in Neural Networks. General observations: Neural Network approaches support dynamic environments [97, 98] and can run fast and conclude, however to achieve this the NN should be pre-trained and this depends on the size of NN, the depth and if the NN is DNN. Thus NN approaches can be considered dynamic and flexible. In order for the NN to be able to tackle all the cases, it is important that the pre-trained step of the NN be executed correctly. For the simple NN category, correct execution can be accomplished when the NN is trained for all cases with a 75% of the data for training and 25% for testing. In case of a non-trained NN, time will be needed for the training and the testing of the NN in order for the NN to be ready to start correctly calculating values from inputs (meanwhile until the NN is trained some errors in calculation will exist). In NN the identification or creation and forming of training and testing data is time consuming. This is a factor that should be considered when NN is used as an intelligent approach. In case that no training or testing data exists, Back propagation NN can be exploited. In this type of NN, the NN is trained by itself by collecting the training data from the environment (e.g., Channel Quality Indicator, Signal To Interference Noise Ratio (SINR), Power, frequencies, etc. of the UE or D2D devices). However, the limitation of this type of NN is that if the NN is not trained enough, some errors in calculation will exist, which in the case of D2D communication may not acceptable. Also, another advantage of the NN is that it can work well with other intelligent approaches (i.e., fuzzy logic) and non-intelligent approaches (i.e., game theory) because it can jointly solve D2D challenges. In addition, the NN approaches in the literature are doing prediction by using some calculated thresholds pre calculated from the control device (BS/D2D) and by using BS data, which is time consuming. Therefore, the NN approaches as they are in the literature cannot run in parallel on D2D devices, thus cannot implement DAI. However, the NN can do better and implement DAI when they are concentrated on the locality of the problem and use information based on a range threshold and handle the problem as local problem.

2.3.3.4 Thompson Sampling and Bayesian Control Group

The D2D communication approaches that use Thompson sampling and Bayesian control mainly focus on the Power Control, Radio Resource allocation, and QoS. General observations: Thompson sampling and Bayesian control (TB) approaches are dynamic as they can easily handle dynamic situations (i.e., like a D2D UE location change). Any changes are handled by forcing recalculation on a generation of action and a response of positive/negative reward in each change by using the identified/known utility. However, TB approaches are not flexible, as they cannot adapt fast to changes in the topology of the D2D Network. This is because the algorithms must rerun from the beginning (with the steps of initialization and action-reward through the maximum utility) so as for the maximum utility to be recalculated, which is a time-consuming process. An advantage of Thompson sampling and Bayesian control approaches is that they can utilize agents. An agent can learn from properties (i.e., frequency, power) and adapt its behavior dynamically [101] in order to solve a problem. This characteristic is beneficial in addressing D2D challenges, which are related to a dynamic environment. Moreover, TB approaches can utilize any other intelligent approach (i.e., fuzzy logic) in its utility function and help in addressing more D2D challenges (e.g., Handover with fuzzy logic). Additionally, even though the group is using distributed control, the group approaches found in the open literature did not implement DAI, as they depended on the calculation of thresholds or limits (like bit rate) in advance. However, these may be high-level directives/set points for the guidance/operation of the devices. If the thresholds are not set up at any particular period they cannot operate independently because the decisions will take are critical for the whole convergence of the algorithm as they vary in the decision process. Therefore, thresholds need to be setup correctly in order to do a successful run. Because the control used in the group approaches is not multilevel control where in this type of control the approach can use the thresholds to in another level of control. Also TB approaches, due to trial and fail, takes time to conclude. Therefore, the re-run execution is considered expensive, as there is an extensive use of CPU, battery and network bandwidth.

2.3.3.5 Evolutionary Algorithms Group

The D2D communication approaches that use Evolutionary Algorithms mainly focus on the technical challenges related to Mode Selection and Radio Resource Allocation. Note that not many papers use EA (only one paper using EA was found and included in this group). General observations: EA approaches are dynamic as they can easily handle dynamic situations (i.e., like a D2D UE location change) by forcing recalculation base on the new location and by using the fitness function to select the best solution and therefore conclude quickly and easily. However, they are not flexible, as they cannot adapt fast to changes in the topology of the D2D Network. More specifically, when a new UE enters/leaves the D2D Network, the algorithms must rerun in order to recalculate the fitness function, crossover and mutation that will be considered for the estimation of the desired frequency, power, and access point to connect. This is considered as a disadvantage of this group since with the rerun of the algorithm there is an extensive use of CPU, battery, and network bandwidth (due to signalling exchange), which may become a prohibitive factor for using EA for the implementation of any D2D solution. Additionally, the execution of EA approaches, due to trial and error, takes time to conclude. Moreover, because the group is using Centralized control, the group approaches found in the open literature could not implement DAI. That is why EA approaches are not so popular in addressing D2D technical challenges. However, EA approaches can use any other intelligent approach (i.e., NN) in its utility function and help in addressing more D2D challenges (e.g., Interference Management with NN).

2.3.3.6 Genetic Algorithms Group

The D2D communication approaches that use Genetic Algorithms mainly focus on the Radio Resource allocation, and to a lesser extent also address D2D challenges related to: Interference Management, Power control and QoS. General observations: Genetic Algorithm approaches are dynamic as they can easily handle dynamic situations (i.e., like a D2D UE location change) by forcing recalculation base on the new location and by using the fitness function to select best solution. By doing this, GA approaches conclude quickly and easy. In some cases, they may conclude and stop without finding the optimum solution, when these are based on threshold defining maximum iterations. As the GA approaches must rerun from the beginning in order to recalculate the desired frequency, power and access point to connect, they cannot adapt fast to changes in the topology of the D2D Network. Thus GA approaches cannot be characterized as flexible. More specifically, likewise with EA, when a new UE enters/leaves the D2D Network, the algorithms must rerun in order to recalculate the fitness function, crossover and mutation that will be considered for the estimation of the desired frequency, power, and access point to connect. This is considered as a disadvantage of this group since with the rerun of the algorithm there is an extensive use of CPU, battery and network bandwidth (due to signalling exchange), which may become a prohibiting factor for using GA for the implementation of any D2D solution. Additionally, the execution of GA approaches, due to trial and error, takes time to conclude. Moreover, even if GA, by default, supports autonomous nodes, we could not identify any GA-based D2D approach in literature using the D2D device as autonomous. GA approaches can utilize any other intelligent approach (i.e., Q-Learning) in its utility function and help in addressing more D2D challenges (e.g.,

QoS with Q-Learning). Moreover, even though the group is using distributed control, the group approaches found in the open literature did not implement DAI, as they depend on the calculation of thresholds/constrains or limits (e.g., max bit rate) and max generations threshold in advance, in order to do a successful run.

2.3.3.7 Particle Swarm Optimization Group

The D2D communication papers that use Particle Swarm Optimization mainly focus on the D2D technical challenges related to Interference Management, Radio Resource allocation and QoS and to a lesser extent also address D2D challenges related to: Mode Selection and Power control. Although trends in D2D communication research do not appear to favour PSO, PSO follows Q-Learning in terms of popularity. General observations: Particle Swarm Optimization approaches can handle dynamic situations (i.e., like a D2D UE location change). This is achieved by forcing recalculation based on the new location and by using the pre-calculated particle velocity and position to select best solution to conclude. The above recalculation is executed quickly because PSO approaches are already guided, even before the UE changes position, toward the best-known positions in the search-space. The PSO, using particle's position updates, aims to make the swarm move towards the best solutions, but the result may not be the optimum. PSO approaches cannot be characterized as flexible, because with changes at the topology of the D2D Network, the group approaches cannot adapt fast. More specifically, when a new UE enters/leaves the D2D Network, the algorithm, due to threshold changes, must rerun and recalculate particles, PSO position, and velocity that will be considered for the estimation of the desired frequency, power, and access point to connect. This is considered as a disadvantage of this group since with the rerun of the algorithm there is an extensive use of CPU, battery and network bandwidth (due to signalling exchange), which may become a prohibiting factor for using PSO for the implementation of any D2D solution. Additionally, the execution of PSO approaches, due to trial and error, takes time to conclude. However, an advantage of the PSO approaches is that they can utilize agents in the solution to identify the best position towards the solution. PSO approaches can utilize any other intelligent approach (i.e., fuzzy logic) in its utility function and help in addressing more D2D challenges (e.g., Handover with fuzzy logic). Moreover, even though the group can use distributed control, the group approaches found in the open literature did not implement DAI, as they depend on the calculation of thresholds/constrains or limits (e.g. max bit rate) and max iterations threshold in advance in order to do a successful run.

2.3.3.8 Ant Colony Optimization Group

The D2D communication papers that use of Ant Colony Optimization mainly focus on the QoS, and to a lesser extent also address D2D challenges related to: Radio Resource allocation. General observations: Ant Colony Optimization approaches are dynamic as they can easily handle dynamic situations (i.e., like a D2D UE location change) by forcing recalculation based on the new location and by using the existing pheromone trails to calculate and select the best solution. The above recalculation is executed quickly because ACO approaches, in order to find the final solution, are using agents (i.e., artificial ants) moving through different paths with different parameters representing all possible solutions. The ants, while exploring their environment during the construction of the path, are directed by each other through the pheromone concentration (e.g., overall throughput) to the resources (that is the end of the path). Nevertheless, the result may not be optimized because the ants do not have a global view of the solutions. Thus, due to some thresholds (max number of iterations allowed) they might select the local optimum solution instead of the global optimum. Moreover, ACO approaches cannot be considered as flexible, as they cannot adapt fast to changes in the topology of the D2D Network. More specifically, when a new UE enters/leaves the D2D Network, due to threshold changes, the algorithm must rerun and recalculate the paths that the artificial ants should follow (e.g., by leaving pheromone; note that the pheromone is the direction to the local optimum solution and this could be the global optimum, but because it is a meta-heuristic approach it does not guaranty the global optimum [128, 129]) in order to recalculate the desired frequency, power, and access point to connect. This is considered as a disadvantage of this group since with the re run of the algorithm there is an extensive use of CPU, battery and network bandwidth (due to signalling exchange), that may become a prohibiting factor for using ACO for the implementation of any D2D solution. Additionally, due to random searching of paths and trial and error, ACO approaches take time to conclude. However, some of the papers using ACO approach are fast to conclude on the task by using a more accurate calculation on pheromone (bias). But even with that, the overall understanding is that ACO approaches are slow due to the fact that at first artificial ants will select random paths before concluding in order to find the best path. ACO approaches can utilize any other intelligent approach (i.e., fuzzy logic) in its utility function and help in addressing more D2D challenges (e.g., Handover with fuzzy logic). Additionally, even if group approaches, by default, support autonomous nodes and agents (which is considered as an advantage), we could not identify an ACO approach in the literature which was using the D2D device as autonomous nor using the full potential of agents. Moreover, even though the group is using distributed control, the group approaches found in the open literature did not implement DAI, as they depended on the calculation of thresholds/constrains or limits (e.g. max bit rate) in advance.

2.3.4 TAXONOMY OF GROUPS BASED ON APPROACH USED FOR D2D COMMUNICATION ESTABLISHMENT

In this section, the groups formed, were further put in taxonomy according to the approach used for the establishment of D2D Communication. More specifically, these groups were classified based on: (i) Spectrum utilization (i.e., Inband or Outband) for establishing the D2D communication links, (ii) the way Control is performed (i.e., Centralized, Distributed, Distributed Artificial Intelligence, Semi-distributed) for establishing D2D communication; and (iii) the D2D Transmission Modes allowed (i.e., D2D relay, D2D cluster, D2D multi-hop relay) for D2D communication (see Figure 2.1).

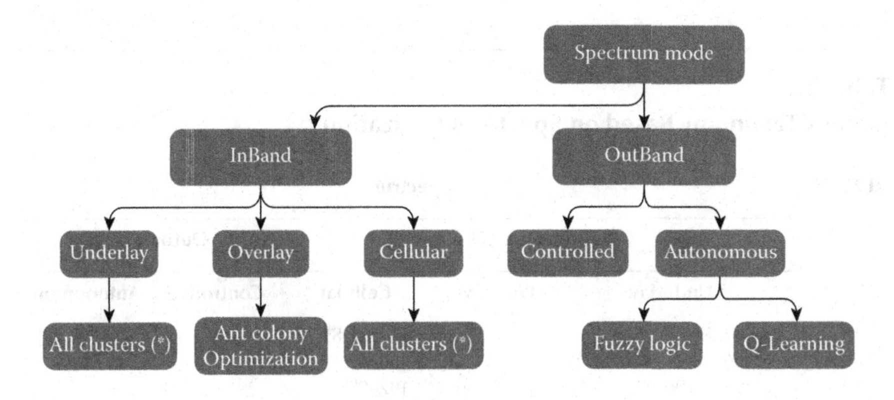

Figure 2.3 Groups taxonomy based on spectrum utilization.

2.3.4.1 Taxonomy Based on Spectrum Utilization

In this section, each group (i.e., intelligent approach) is classified on the frequency perspective (see Figure 2.3 and Table 2.2). More specifically, for each group we examined the following Frequency Mode Types/Spectrum Utilization (as shown in the Section 2.2.2.2.1) :

1. How the spectrum is utilized. Here we checked if the group uses the BS frequencies specified for D2D, if it uses frequencies that are reused, or if it is using frequencies that exist in other technologies (i.e., Wi-Fi, Bluetooth).
2. What type of frequencies are used (Inband or Outband).
   a. Inband D2D [55]: In this type of D2D communication, the cellular spectrum for both D2D and cellular links is used. By using Inband communication, higher control over cellular (i.e., licensed) spectrum is gained as interference is controllable which improves QoS provisioning. We have three types of Inband D2D (see Section 2.2.2.2): (i) Underlay; (ii) Overlay; and (iii) Cellular mode.
   b. Outband D2D [55]: In this type of D2D communication, the D2D links exploit unlicensed spectrum. The motivation behind using Outband D2D communication is to eliminate the interference issue between D2D and cellular link. The disadvantage of the outband D2D is that it has the uncontrolled nature of unlicensed spectrum. It should be noted that only cellular devices with two wireless interfaces (e.g. LTE and Wi-Fi, Bluetooth, Wi-Fi direct) can use Outband D2D, and thus users can have simultaneous D2D and cellular communications. Outband D2D can be established in two modes (see Section 2.2.2.2): (i) controlled mode and (ii) autonomous mode.

For the following reasons we assert that a group (Intelligent approach) should exploit all types of spectrum utilization and be ready to use each one of them for the D2D network implementation:

Table 2.2
Groups Taxonomy Based on Spectrum Utilization

AI/ML IA	Spectrum				
	Inband D2D			Outband D2D	
	Underlay	Overlay	Cellular	Controlled	Autonomous
FL	√ [44, 52, 81]		√ [44, 52, 81]		√ [44, 52]
QL	√ [66–68, 83–86, 88, 89]		√ [66–68, 83–89]		√ [85]
NN	√ [92–94, 96, 97]		√ [92–98]		
TB	√ [101]		√ [101]		
EA	√ [105]		√ [105]		
GA	√ [108, 109]		√ [108, 109]		
PSO	√ [111–118]		√ [111–118]		
ACO	√ [123]	√ [122]	√ [122, 123]		

Inband Overlay: In this type, a rigid fraction of the licensed spectrum is reserved for D2D UEs. This spectrum utilization type is important as one band should be kept for emergency use (Inbound Overlay) when a UE has to communicate due to an insistent (e.g. car accident, ambulance) with special rights.
Inband Underlay: In this type, D2D communications takes place over the same licensed spectrum intended for legacy cellular simultaneously. This spectrum utilization type is important since the D2D devices and other UEs can reuse bands, because the frequencies are limited, and the task of the approach must be to satisfy all UEs even the devices in a cell that is overloaded. Therefore, this is considered a most valuable type of spectrum utilization.
Inband Cellular: This spectrum utilization type is important since D2D must use in some cases its cellular resource to communicate between another D2D devices without interfering with the BS.
Outband Controlled: In this type, D2D UEs exploit unlicensed spectrum to communicate and have access to the BS. This spectrum utilization type is important since the case of sharing a link to BS from a D2D Cluster and act as Cluster Head should be considered, since the internal communication between the D2D UEs will not pass the BS. Consequently, there is a reservation of resources.
Outband Autonomous: In this type, D2D UEs exploit unlicensed spectrum to communicate and they do not have access to the BS. This spectrum utilization type is important since the case of a D2D relay node sharing a link to Wi-Fi Access Point (AP) or any other Access Point different than the BS, should be considered. Therefore, the total sum rate in the network increases.

However, as shown in Figure 2.3 and Table 2.2 above, none of the Intelligent Approaches (groups) implements all of the features. More specifically:

ACO is the only group that implements all modes of Inband D2D (Undelay, Overlay and Cellular). All other groups implement only Underlay and Cellular.
FL and QL are the only groups that implement Outband D2D, however this only for Autonomous mode.
None of the groups implement Outbound Controlled.

2.3.4.2 Taxonomy Based on D2D Transmission Mode Solutions Allowed

In this section, each group (i.e., intelligent approach) is put in taxonomy based on the D2D Transmission modes (i.e., D2D relay, D2D cluster, D2D multi hop relay, D2D Direct) allowed for D2D Communication (see Figure 2.7 and Table 2.4). More precisely, by examining the Transmission mode we have the following (as shown in the Section 2.2.2.2.2):

If D2DMHR is supported it means that this approach can have optimized paths and the approach can have minimum costs on transmission.
If D2DSHR is supported it means that the approach can have connection with the internet (external network) at the same time with the interchange of data.
If D2D Cluster is supported, it means that the approach can have a small adhock "network" under the network of BS, with a D2DSHR acting as CH (Cluster Head) and D2D devices under the CH (D2D devices under the same cluster) to interchange data between each other without affecting the BS.

The transmission architectures of a D2D base communication illustrating how they form relation with other nodes is shown in Figures 2.4 and 2.5, and further explained below:

D2D Relaying: In this transmission mode, a D2D device forms a Link Share of bandwidth between BS/UE and other UE(s) Devices. The share bandwidth could be directly connected to a BS (or other UEs (that could be also in D2D relay mode) or another Access point. Because 5G cellular networks enable using direct communication between devices as a relay strategy for coverage extension the D2D relay can be established. In D2D Relay (D2DSHR) both backhaul and D2D transmissions are performed in uplink cellular resources, and are subject to cellular uplink power control. The relay selection and resource allocation is a problem to solve for D2D-relaying in a multi-user, multi-carrier, and multi-cellular network [39]. The technology that can be used in order to form D2D Relay (D2DSHR) is LTE Direct and Wi-Fi Direct [130].
D2D Multi hop Relay (D2DMHR) is a sub-type of D2D Relay: In this transmission mode, a D2D device forms a Link Share of bandwidth between D2D Relay/BS and other D2D Relay devices (so both backhaul and D2D transmissions are performed in an uplink with other D2D relay node as a bridge and they are subject to the other D2D relay node control). The use of D2D multi hop Relay addresses the communication needs of UEs inside mobile network coverage, and those UEs that suffer from scarce radio coverage [40]. The technology that can be used in order to form D2D Multi hop Relay is LTE Direct and Wi-Fi Direct [130].

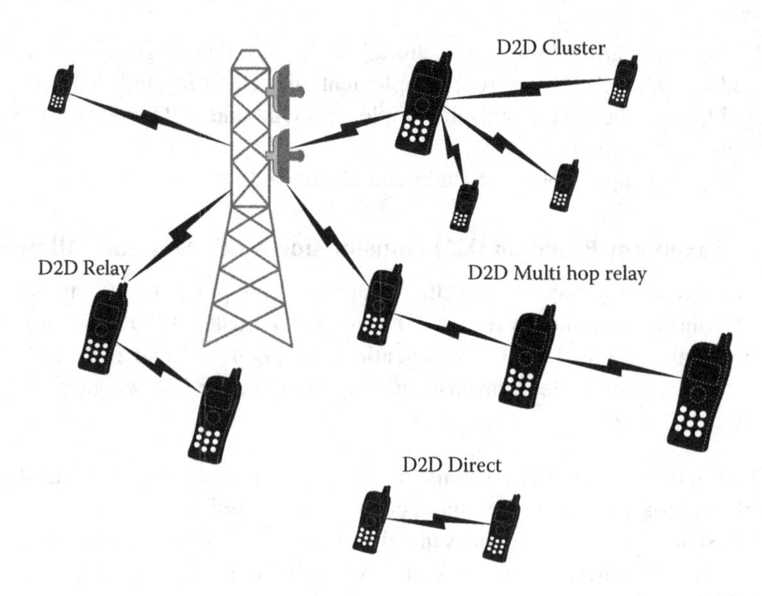

Figure 2.4 D2D transmission modes.

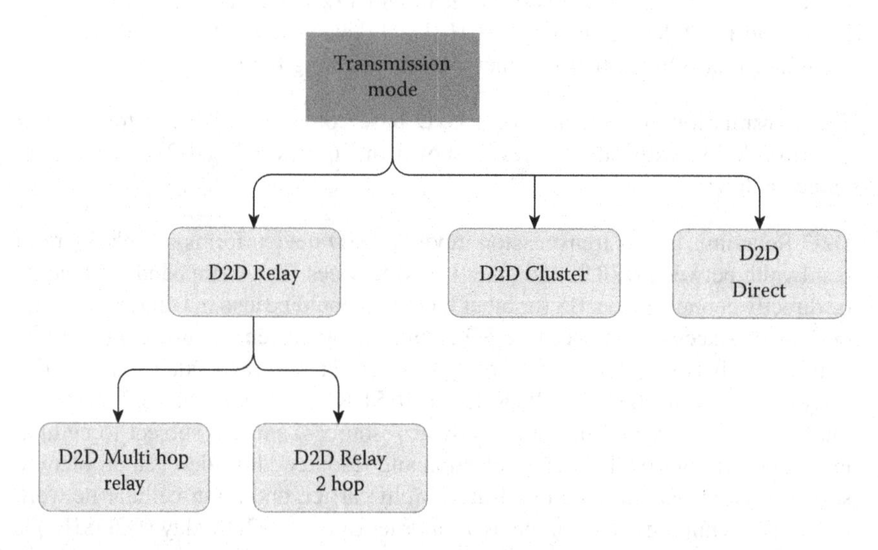

Figure 2.5 Groups taxonomy based on transmission mode.

D2D Cluster (D2D LAN [41]): In this transmission mode, a D2D device(s) connects to a D2D relay device for accessing the network and if the devices are more than one they can intercommunicate between each other through the common D2D Relay (D2DSHR) device. The D2D Relay (D2DSHR) device is called cluster Head and it forms a Link Share of bandwidth between BS/D2D Relay/D2D multi hop relay device and other D2D devices under it. The clustering concept offers features that can utilize direct communication in a cellular network in order to keep local communication between D2D devices in the same cluster local. In addition, the traffic between communicating devices if routed via the core network it increases the network load, data delay and base station resource utilization. In D2D cluster, concept devices can be assigned to direct communication mode utilizing cellular network resources. Direct communication mode excludes the unnecessary core network involvement and enhances the base station resource utilization. In D2D cluster, there exists a Cluster Head (CH) which utilizes D2D Relay (D2DSHR) Node Transmission mode. If the CH is using Inband (Overlay/Underlay) or Outband Controlled then the cluster has access to the BS, else it must use outbound-D2D autonomous mode in order to access the network [131]. The technology that can be used in order to form D2D Cluster is LTE Direct and Wi-Fi Direct. Wi-Fi Direct is already mature enough to form clusters with CHs [42, 130].

D2D Direct: In this transmission mode, two UEs connect to each other by using licensed or unlicensed spectrum. The two D2D UEs only communicate with each other (also called Full-Duplex D2D). The technology that can be used in order to form D2D Direct is LTE Direct and Wi-Fi Direct.

Note that for the Taxonomy of Transmission we have added an extra category. The category is the D2D Relay of 2 Hops because there are approaches that define that they use such a connection by restricting the depth of the path to only two.

Given above, we ascertain that a group (Intelligent approach) should strive to implement all D2D Transmission modes for the following reasons. With D2DMHR the intelligent approach can expand to areas that cells cannot support or handle overload situations in a cell. In addition, with D2D Relay transmission mode the intelligent approach can support HetNets and expand network coverage. Likewise, with the D2D Cluster, the approach can save bands, bandwidth usage and increase sum rate. D2D Direct is by default the mode that D2D communications support, so all intelligent approaches by default should support this mode.

However, as shown in Table 2.3 above, none of the Intelligent Approaches (groups) supports all the D2D Transmission modes and moreover, none of them support D2DMHR in more than 2 depths. More specifically:

- FL, GA, and PSO supports D2DSHR, D2D Cluster and D2D Direct Transmission modes
- QL and ACO, additionally with the aforesaid, are the only that support D2DMHR 2 Hop.
- NN, TB and EA supports only D2D Direct.
- None of the groups support D2DMHR.

Table 2.3
Groups Taxonomy Based on Transmission Mode

AI/ML IA	Transmission Mode				
	D2D Relay	D2D Multi-hop Relay 2 Hop	D2D Multi Hop Relay	D2D Cluster	D2D Direct
FL	√ [44, 52, 81]			√ [44, 52, 81]	√ [44, 52, 81]
QL	√ [85–87]	√ [85–87]		√ [85–87]	√ [66–68, 83–89]
NN					√ [92–98]
TB					√[101]
EA					√ [105]
GA	√ [108]			√ [108]	√ [108, 109]
PSO	√ [111, 116]			√ [116]	√ [111–118]
ACO	√ [122]	√ [122]		√ [122]	√ [122, 123]

2.3.4.3 Taxonomy Based on Control Performed for D2D Communication Establishment

In this section, each group (i.e., intelligent approach) is put in taxonomy based on the way Control is performed (i.e., Centralized, Distributed, Distributed Artificial Intelligence, Semi-distributed) in a Device for establishing D2D communication (see Figure 2.7 and Table 2.4). This taxonomy was considered important, as there are certain disadvantages/advantages of a control type over other control type that is performing D2D communications. More precisely, the types of control (identified from [37]) that can be used for the establishment of D2D Communication links, are categorized as follows (see Figure 2.6):

- Centralized: The BS completely oversees all the UEs (regular and D2D), and operates as the central controller responsible for managing interference/connections/path establishment, etc., in the cell.
- Distributed: The procedures of managing interference/connections/path establishment, etc., in the cell, is performed autonomously by the UEs themselves. This scheme reduces the control and computational overhead and is particularly appropriate for large size D2D networks.
- Distributed AI (DAI): A separate case of distributed AI control where all control processes performed by the UEs can begin asynchronously and run in parallel in a distributed manner.
- Semi distributed (Hybrid): The procedures of managing interference/ connections/ path establishment, etc., in the cell, are performed by the BS (Centralized) and the UEs (Distributed) in collaboration. The aim is to adopt the strong points of each approach for better performance.

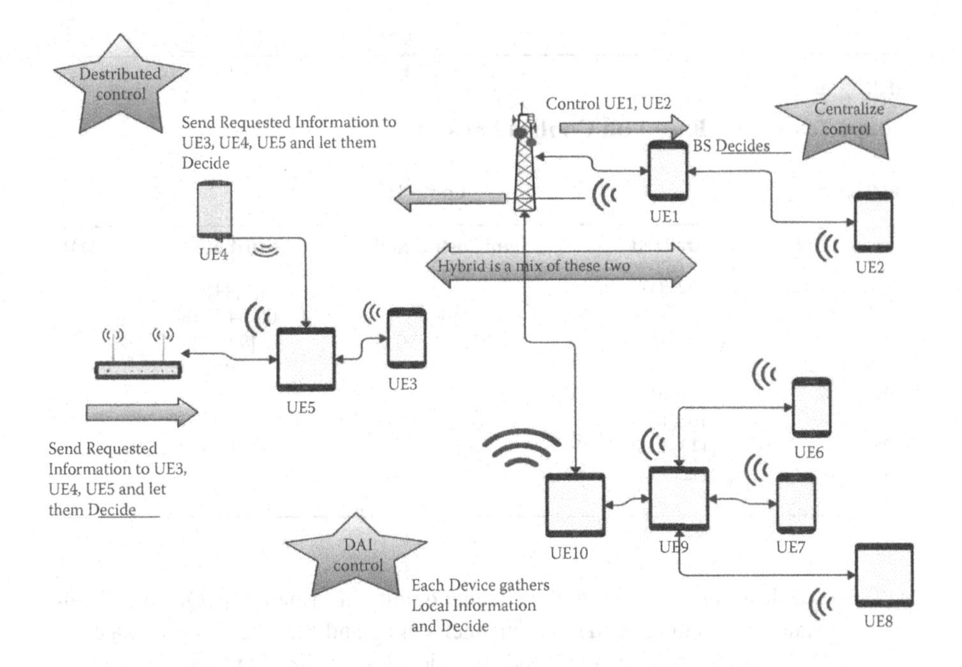

Figure 2.6 Types of control in D2D communication: centralized, distributed, DAI, hybrid.

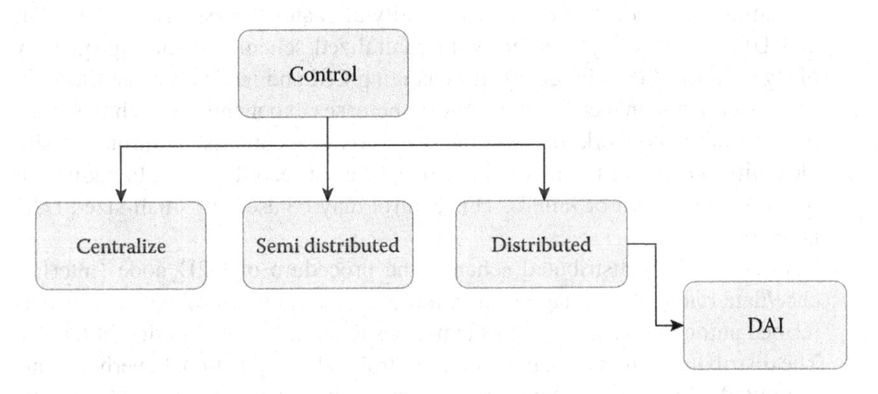

Figure 2.7 Groups taxonomy based on control performed.

Therefore, this classification is based on who controls the whole process. In the following section we will examine how each control type controls D2D communication and in more depth the mechanisms of each type of control:

- Centralized: Within the centralized technique, the D2D nodes are managed by the eNB (maybe a different entity than eNB could also do the control). The controller manages, among others, interference, connections, path between Cellular UEs (CUEs) and D2D UEs (DUEs). The BS collects information from

Table 2.4
Groups Taxonomy Based on Control Performed

AI/ML IA	Control			
	Centralized	Semi-Distributed	Distributed	DAI
FL	√ [52, 81]		√ [44]	
QL	√ [66, 67, 83, 87, 89]	√ [86]	√ [68, 84, 85, 88]	
NN	√ [93, 95]	√ [92, 96, 98]	√ [94, 97]	
TB			√ [101]	
EA	√ [105]			
GA	√ [108, 109]			
PSO	√ [113–118]		√ [111, 112]	
ACO	√ [122, 123]			

the wireless network, as e.g. the channel quality information (CQI), the Channel state information (CSI), the channel status, and the interference stage for each UE within the network, and then decides on the channels to assign to every UE with the proper format and power level. Primarily based on the information received, the authoritative entity allocates the assets to every CUE or DUE. The primary problem with centralized schemes is the big quantity of signalling overhead required for changing CSI and feedback from the UEs. Moreover, the management complexity increases exponentially with the range of users in the network, because the operation is accomplished through a single entity, which has to process large quantities of records (data). In addition it poses a single point of failure. This control may be used for small-sized D2D networks.

- Distributed: In a distributed scheme, the procedure of D2D node (interference/data rate/path) management in not executed on a central entity; it is performed autonomously by DUEs themselves without the intervention of the BS. The distributed scheme decreases the control and computational overhead, due to limited CSI (channel state information) exchange and due to reduced message exchange. However, in this scheme, facilitating and handling interference is more difficult than the centralized case. Nevertheless, this approach may be considered in all ranges of networks (small, medium, large).
    - Distributed Artificial Intelligence (DAI): This is a category of distributed control scheme which solves complex learning, planning, and decision-making problems. Additionally with the Distributed Control described above, this DAI scheme supports perfectly parallel workload.[5] More specifically, tasks with parallel control are performed by all D2D devices in the network. Thus DAI is able to exploit large-scale computation and spatial distribution of computing resources and the control is done by

each node in parallel. The intelligent agent approaches can only support this type of control. Moreover, this type of control can be considered for all ranges of networks (small, medium, and large).

- Semi-Distributed: In spite of the fact that both centralized and distributed schemes have their good points and drawbacks, tradeoffs can be accomplished between them. Such D2D (for interference/data rate/path) management schemes are the "semi-distributed" or "hybrid". Within the semi-distributed (for interference/data rate/path) management schemes, different levels of involvement can be defined. Control is done together by D2D devices and eNB. Such schemes could be usefully adopted for the medium range of networks.

Given above, the authors assert that a group (Intelligent approach) should consider the Distributed Artificial Intelligence (DAI) control, as it can address many of the open issues. More specifically, Distributed Artificial Intelligence (DAI) is an approach which can solve complex learning, planning, and decision-making problems. It is embarrassingly parallel, thus able to exploit large scale computation and spatial distribution of computing resources. That is little or no effort is needed to separate the problem into a number of parallel tasks. This is often the case where there is little or no dependency or need for communication between those parallel tasks, or for results between them [17]. These properties allow it to solve problems that require the processing of very large data sets. DAI systems consist of autonomous learning processing nodes (agents), which are distributed, often at a very large scale. DAI nodes can act independently and partial solutions are integrated by communication between nodes, often asynchronously. By virtue of their scale, DAI systems are robust and elastic, and by necessity, loosely coupled. Furthermore, DAI systems are built to be adaptive to changes in the problem definition or underlying data sets due to the scale and difficulty in redeployment [132].

However, as shown in Figure 2.7 and Table 2.4 above, none of the Intelligent Approaches (groups) implements DAI distributed control, even if they can support it. More specifically:

- QL and NN implements all the types of Control except DAI.
- EA, GA and ACO implements only Centralized Control
- TB implements only Distributed Control
- FL and PSO implements Centralized and Distributed Control

2.3.5 COMPARATIVE ANALYSIS OF THE DIFFERENT GROUPS

Prior to a detailed analysis of the papers identified during the collection of related state-of-the-art work, some highlights are presented next. The aim is to provide to the readers, through comparison tables and graphs, an overall idea of the outcomes extracted from this research regarding: (i) the groups formed; (ii) the popularity of each Intelligent approach (group) used for addressing D2D Challenges as well as the trends in research throughout the years (from 2010 to 2019); (iii) the D2D challenges addressed by each group; (iv) the D2D challenges that still remain as an open issue for Intelligent Approaches; and (v) Features that are considered important to be supported by the groups formed.

More specifically, as illustrated in Figure 2.8 below, a total of 8 groups have been formed, one for each intelligent approach considered in our analysis. These are FL, Q-Learning, Neural Networks, Thompson Sampling and Bayesian Control, Evolutionary Algorithms, Genetic Algorithms, Particle Swarm Optimization and Ant Colony Optimization. Additionally in this figure, the papers included in each group as well as the total citations credited (up to 9/10/2019), are shown. Based on these groups, relevant papers have been collected, analyzed and grouped based on the intelligent approach they exploited to address a D2D Challenge.

The popularity of each approach and the trends in research throughout the years 2010 - 2019 is provided in Table 2.5. By using the numbers of papers as a metric, the most popular throughout all years is the Q-Learning approach. This was expected as most of the papers that exploits Q-Learning try to address the D2D challenges in the perspective on real-time evaluation of action-reward. In Q-Learning, each agent can resolve maximization problems efficiently, if the reward function is defined correctly. In addition the Q-Learning approach implements a Q table and can keep history of decisions, therefore is more flexible than other approaches. Regarding popularity, Particle Swarm Optimization follows Q-Learning probably due to the way it works. Precisely, PSO works as Optimization AI approach for finding the optimal solution (e.g., sum rate). However, in terms of citations reflecting the overall academic acceptance of the approach, PSO papers comes first with a total of 127 citations, followed by Q-Learning and Tomson Sampling and Bayesian Control with 28 citations.

Third and fourth in line in popularity is Neural Networks and FL. It is worth noting that Fuzzy logic solves multiple D2D challenges and therefore the papers of FL offer studies covering most of the D2D challenges. Furthermore, it was observed that no approach/paper offers solution proposals to cover all the spectrum of the D2D Challenges. However, recent trends in research appear to be Q-Learning and Neural Networks (perhaps capitalizing on the current popularity of AI and deep learning) for addressing D2D challenges as most articles in 2018/2019 adopt these approaches.

For quick reference, Table 2.6 and Figure 2.9 summarizes the D2D challenges addressed by each group as well as the D2D challenges that we identified still remain as open issues. More specifically Device Discovery, Mode Selection, Security of D2D Communication, Cell Densification and Offloading, D2D using mmWave Communication and Handover of D2D device are challenges that still need further research on how Intelligent Approaches can be adopted to address these. Therefore there are opportunities for use of AI techniques in the above challenges.

Features that are considered important to be supported by the groups formed in further improving the D2D communication are the following:

1. Dynamic Implementation (DI): The intelligent Approach should take into consideration a dynamically changing environment, where the UE location is changing rapidly and therefore the band selection and power calculation (in Underlay) need to be updated during the next time period. Also, for any small change in mobile network, the algorithms must not rerun. In addition, the dynamic implementation

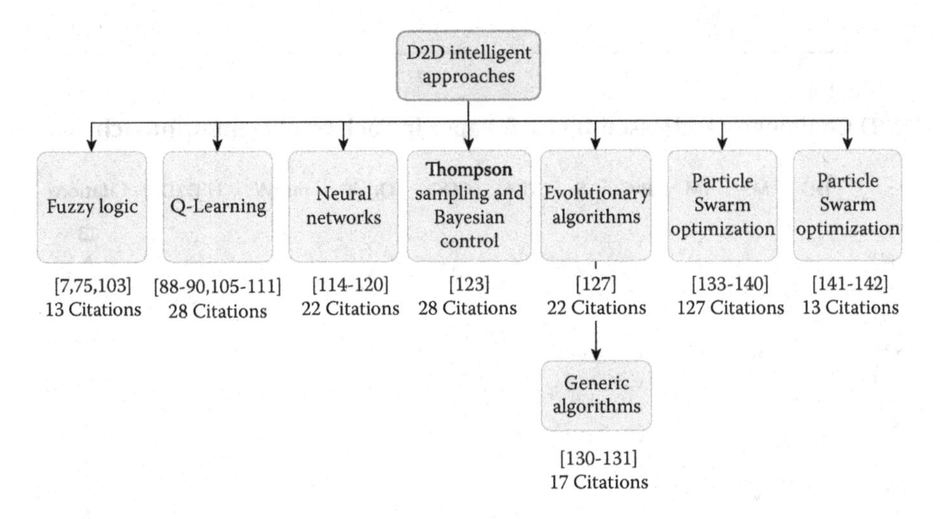

Figure 2.8 Groups formed and papers included.

Table 2.5
Popularity of Each Approach and the Trends in Research 2010–2019

	Year of Publication										
	10	**11**	**12**	**13**	**14**	**15**	**16**	**17**	**18**	**19**	**Total**
FL								1	2		**3**
QL									7	3	**10**
NN									5	2	**7**
TB						1					**1**
EA				1							**1**
GA					1			1			**2**
PSO		1		1	2	1		1	1	1	**8**
ACO								1	1		**2**

should be able to handle new dynamic situations (i.e., adding new technologies, as e.g. a new Device that supports Bluetooth sharing). All groups are dynamic.

2. Multi-Cell environment consideration (Multi-Cell): With this feature supported, the interference of the neighboring cells can be better handled. Thus interference can be better controlled resulting in better spectral efficiency.
3. HetNet support (HetNet): With this feature supported both cellular and other Radio Access Technologies (Heterogeneous networks) that will have different protocols support like Wi-Fi, Bluetooth, ZigBee, Lora, 3G, 4G, etc., can be considered in the D2D link establishment.

Table 2.6
D2D Challenges Addressed by Each Paper in Each Intelligent Approach

	DD	MS	IM	P-C	S	RRA	CDO	Qos'P	mmW	H'D2D	Citations
FL	√		√	√	√	√	√	√		√	**13**
[44]	√		√		√	√	√	√		√	6
[52]	√		√	√		√		√		√	5
[81]				√						√	2
QL		√	√	√	√	√		√			**28**
[66]				√		√		√			6
[83]				√				√			5
[84]				√		√		√			8
[85]				√				√			0
[86]		√	√	√				√			3
[67]						√		√			1
[87]				√				√			2
[68]			√			√		√			1
[88]				√	√			√			1
[89]				√				√			1
NN			√	√		√		√	√	√	**22**
[92]			√	√							0
[93]			√	√							4
[94]			√	√							4
[95]			√	√		√					8
[96]			√	√							5
[97]			√			√		√			0
[98]				√				√	√	√	1
TB				√		√		√			**28**
[101]				√		√		√			28
EA		√				√					**22**
[105]		√				√					22
GA			√	√		√		√			**17**
[108]				√		√		√			13
[109]			√			√					4
PSO		√	√	√		√		√			**127**
[111]								√			0
[112]			√	√				√			22
[113]		√	√			√					74
[114]		√	√			√					12
[115]			√			√		√			6
[116]			√			√		√			5
[117]			√			√		√			5
[118]			√	√		√		√			3
ACO						√		√			**13**
[122]								√			10
[123]						√		√			3

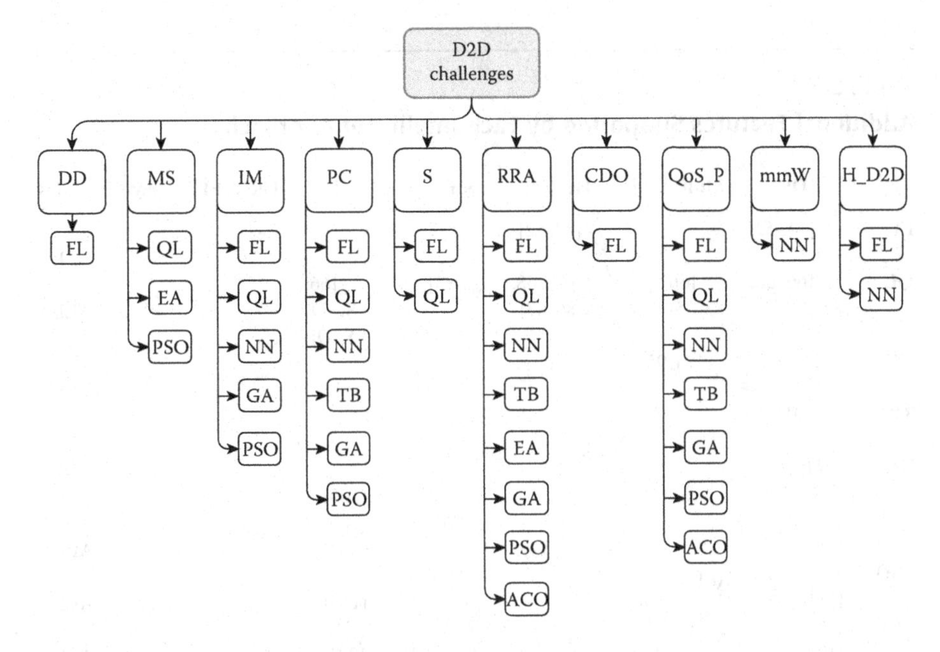

Figure 2.9 D2D challenges addressed by each intelligent approach.

4. QoE support (QoE): With this feature supported, the Quality of Experience of the User in the Network is considered, thus guaranteeing at least the minimum desired data rate of the user demand.
5. Fairness in UEs bandwidth usage (F): With this feature supported, the fairness factor is considered as the fair handling of D2D devices among the whole network coverage (i.e. UEs closed to BS and far from BS), thus guarantying the minimum measurement factor like data rate for all the UEs in the network.
6. Ultra-reliable low latency (URLL): With this feature supported, the requirements regarding the low latency and ultra-high reliability of the D2D communication link are considered [133].
7. Fault Tolerance (FT): is the property that enables a system to continue operating properly in the event of the failure of some of its components (or one or more faults within) [134].

Based on the analysis performed, Table 2.7 indicates which of those features are supported by each intelligent approach. As shown in the table below, Q-Learning is the only one that addressed all except URLL, FT and mMTC.Additionally, all of the groups can support the eMBB (as discussed in Section 1) use case because they focus on the improvement of the sum rate of the network. However, they can not support the mMTC (as shown in Section 1) use case because the simulations executed by the approaches are with a small number of devices under the D2D network.

Table 2.7
Additional Features Supported by Each Intelligent Approach

	DI	Multi-Cell	HetNet	QoE	F	URLL FT	mMTC	eMBB
FL	√ [44, 52, 81]		√ [44, 52]					√ ALL
QL	√ [66–68, 83–89]	√ [66, 68, 83, 85]	√ [66, 68, 84, 85, 87]	√ [83]	√ [66, 68, 84, 85, 89]			√ ALL
NN	√ [92–98]	√ [94, 95]						√ ALL
TB	√ [101]			√ [101]				√ ALL
EA	√ [105]				√ [105]			√ ALL
GA	√ [108, 109]			√ [108]	√[108]			√ ALL
PSO	√ [111–118]	√ [114]		√ [112]	√ [115]			√ ALL
ACO	√ [122, 123]	√ [122]		√ [122, 123]	√[122, 123]			√ ALL

2.3.6 CONCLUDING REMARKS, OBSERVATIONS AND OPEN ISSUES

The aim of this section is to provide concluding remarks based on the examined literature and identify any pending open issues and challenges that the approaches, in our opinion, did not address, fully or otherwise. Moreover, general observations highlighting some open issues/weaknesses in the existing literature have been identified, that would benefit by further investigation.

Ideally, D2D communications should be a problem solved by the devices that want to communicate in a D2D manner. Therefore, it must be seen as a local problem (i.e., only between the proximate D2D devices) and not a global problem (i.e., taking into consideration all D2D devices in the Network). Hence, not the BS but the D2D device should ideally handle the control, support and security. From the global view perspective, the D2D devices should increase total sum rate, reuse frequencies, create clusters for sharing bandwidth and provide disaster recovery mechanisms to the network in order to contribute to the effective operation of the network.

Any proposed solution should seek to be intelligent for the following reasons: (i) Reinforcement learning is an important aspect for self-organizing networks; (ii) DAI control is only implemented by using agent-based intelligent approaches; (iii) Intelligent approaches can analyze more and deeper data like the frequencies around the D2D device and what appropriate frequency and power can be used in the case of Underlay D2D communication; (iv) Intelligent approaches achieve increased accuracy (i.e., image classification and object recognition); (v) Intelligent approaches can implement an autonomous, flexible and dynamic system; (vi) Intelligent approaches can respond quickly in emergency situations like disaster recovery of a network; and vii) Intelligent approaches can jointly solve some of the challenges. Therefore, the intelligent approaches are expected to offer effective solutions in the implementation of D2D technical challenges.

For the intelligent groups analyzed, a vast number of intelligent approaches in the groups in order to implement D2D communication they necessitate hardware changes at either BS or UE or both. This necessitates the intervention of telecom companies and mobile manufactures, to support the approaches. Also, as shown in the analysis provided above, most of the intelligent groups are dynamic. However, as most of them are not flexible and slow in execution, the Mobile Network may have timeout during: (i) handover; (ii) connection establishment to mobile internet; and (iii) connection establishment to mobile network. Furthermore, even if some of the intelligent approaches (by default) can support autonomous devices, the considered papers did not utilize this characteristic in the D2D intelligent approaches.

Most intelligent approaches are used as utility function/basic measurement value the following: (i) Data Rate; (ii) Sum Rate; (iii) SINR; (iv) SNR; (v) Power (Manipulation on power of D2D in order to reduce interference); (vi) QoE/QoS (Power of battery of device/min data rate); (vii) Spectral efficiency; (viii) Weighted sum rate (of all D2D) and (ix) Location(distance). The aforesaid metrics are important for the implementation of the D2D challenges and an intelligent approach should consider these in the implementation of the solution. However, some of the metrics are connected mathematically through formulation by each other (i.e., if the approach has better Data Rate then it also has better Sum Rate). In addition, it is important for new metrics to be introduced by the intelligent approaches, metrics that will be used by autonomous and distributed intelligent applications.

The ACO and PSO try to form the problem as a maximization problem and solve it. This is achieved by solving an equation that locates optimal solutions by moving through a parameter space representing all possible solutions. The most important aspect in these approaches is the correct formulation and implementation of the utility function, which is a good thing and it should be example for other approaches. The intelligent approaches should be adaptable enough in order to be used jointly or as supporting any other intelligent approach and hence solve the D2D challenges. Based on the finding of this survey conducted, there are no intelligent approach proposals that jointly solve all D2D Challenges. It is our book that will show that the joint solution of D2D challenges should be a feasible goal if the approaches implement DAI with a framework that will jointly use multiple intelligent approaches. In this case, each D2D challenge can adopt the most appropriate intelligent approach (e.g. fuzzy logic with handover, Neural Networks with interference management) and the cross side effects will be handled by the framework that it will tackle the D2D challenges.

Below, we provide a summary of some general observations, identified during our analysis, highlighting some open issues/weaknesses of the existing literature that would benefit by further investigation:

- An approach that solves the joint implementation of all D2D technical challenges is lacking.
  Distributed Artificial Intelligence (DAI) implementation[6] of intelligent approaches is lacking. In D2D the global problem can be separated to pieces of local small problems (locality of small D2D Clusters) and solved by using DAI and true distribution on local level. This is expected to be a powerful solution to the D2D challenges.
- An intelligent approach, which proposes an autonomous solution without the use of the global network data, for pre calculation of thresholds, does not exist in the literature.
- Even though D2D is a locality issue (i.e., only between the proximate D2D devices) most of the approaches handle it as a global issue (at the BS). Therefore, they do not use only data (i.e., SNIR, CQI, Power, frequencies, etc.,) used by the D2D and UE Devices in their proximity but data related to all the D2D and UE Devices in the Network, which are stored in the BS. These requests of data from the BS, create many exchanges of messages, which may cause excessive signalling overhead.
- A vast number of approaches need hardware change at the BS and the UEs that is expensive and difficult task to do.
- In order to be flexible the approach should use modularity in implementation in order to change/add easily the major components (i.e., Telecommunication Interfaces, Communication Protocols, etc.). Only the intelligent approaches that used Fuzzy logic and NN, are considered as flexible.
- Based on the low number of citations (270) on Intelligent D2D approaches found in the open literature, encourages of the Intelligent Approaches community to turn its attention to addresses these D2D challenges.
- Large opportunities for using AI techniques in the following D2D Challenges: (i) Handover of D2D device; (ii) Device Discovery; and (iii) Security of D2D Communication.
- Inadequate research is performed on D2D Intelligent approaches using the following: (i) Multi-hop relay D2D (more than two hops); (ii) Dynamic networks; and (iii) Flexible networks.
- There is no work that supports self-organizing networks[7] (all three categories: Self-configuration, Self-optimization and Self-healing).
- Not a lot of approaches make use of HETNETS.
- Not a lot of implementations are flexible enough in HETNETS to support a variety of other interface technologies (for example mmWave) in D2D Communications. The mobile interface technologies like mmWave could be handled as modules in a modular implementation.
- Only FL adequately addresses the D2D technical challenge regarding Cell Densification and offloading (as the simulations shown in the literature).
- Only FL addresses the D2D technical challenge regarding Device Discovery.
- Only Neural Networks adequately address the D2D Challenge regarding the usage of mmWaves in D2D Communication.

- Only FL and Neural Networks address the D2D technical challenge regarding handover of D2D device.
- Only FL and Q-Learning effectively address the D2D technical challenge regarding Security of D2D Communication.
- There is no implementation that supports the D2DMHR with more than two hops in depth.
- Not a lot of papers support edge computing.
- An intelligent approach utilizing all spectrum modes is lacking. More specifically, Outband (Controlled/Autonomous) is not used by many groups as an alternative gateway to web access.
- an intelligent approach utilizing all spectrum utilization methods is lacking.
- An intelligent approach utilizing all transmission modes is lacking.
- An intelligent approach handling Ultra-Reliable Low Latency (URLL) feature in D2D communication is lacking. The advantages on URLL are examined in some papers [133].
- Not any investigated approach supports Fault Tolerance.

Based on the outcomes and the discussion provided above, in Table 2.8 we identify the intelligent approaches that are most suitable to be used for addressing specific D2D challenges in terms of time, messaging, speed and computation. Therefore , in this part of the section we will state some key observations for AI/ML and D2D for 5G Wireless Systems and we will propose a road-map in order to tackle D2D Challenges at 5G communication.

Due to the complexity of the D2D Challenges, Artificial Intelligence (AI)/Machine Learning (ML) based techniques, thanks to their learning, classifying and controlling capabilities, can be employed to facilitate solving the D2D Challenges in a more efficient manner. In addition, they are widely utilized for maximiza-

Table 2.8
Intelligent Approaches Suitable for Addressing Specific D2D Challenges

AI/ML IA	Challenges									
	DD	MS	IM	P-C	S	RRA	CDO	QoS'P	mmW	H'D2D
FL	√				√		√	√		√
EA								√		
GA								√		
PSO		√						√		
ACO								√		
FL								√		
QL			√	√		√		√		
NN			√	√		√		√	√	

tion/optimization/categorization problems, which makes them perfect candidates for solving the D2D challenges [48]. The aim of utilizing AI/ML, is to allow: (i) autonomous decentralized control; and (ii) collaboration in collection, sharing, and forwarding information in a multihop manner. In addition, AI/ML has the capability to gather relevant information in real time. This is considered a key to leveraging the value of the D2D as such information will be transformed into intelligence which will facilitate the formation of an intelligent environment [137].

AI/ML can be used in order to address jointly all the D2D challenges by implementing a distributed autonomous control environment (i.e., DAI or Distributed Machine Learning (DML)), since as specified above, D2D is a local and not a global problem. A local view of the problem could also aim to facilitate the implementation of the D2DMHR with more than two hops in depth and security of D2D Communication in an efficient manner. Also, by exploiting their learning capabilities (Reinforcement Learning (RL[8]) or Simple Learning (SL)) of each intelligent approach a more optimized Interference Management, Radio Resource Allocation, guarantee QoE and QoS and Power Control can be implemented. Additionally, by intelligently building on the historical information, a more optimized routing path can be selected by the D2D UE improving thus the required QoS. Moreover, an AI/ML technique whenever practical, with the use of RL they are more successful on acting on an unexpected network event (i.e. BS has power cut), in the purpose of realizing a dynamic and flexible D2D communication adapted to the dynamic and flexible nature of Mobile Networks. Furthermore, AI/ML must utilize the whole spectrum (Inband and Outband) and all transmission modes, so as to increase spectrum efficiency and conserve energy. In addition, Self-Organizing Networks (SON) adaptability in a dynamic Mobile Networks environment is very important. Thus, the solution provided by AI/ML approaches must be adaptable. Thus the above guidelines for the adoption and design of AI/ML approaches for D2D we expect would allow one to provide an effective D2D solution, in its totality,and hence contribute toward the achievement of the ambitious guidelines set out for 5G .

Overall, based on the observations extracted from road-map, we identify that there are still opportunities, and a need, for using AI/ML techniques for addressing D2D Challenges. Specifically, inadequate research is performed on D2D Challenges related to Handover of D2D device, Device Discovery, Security of D2D Communication, Cell Densification and Offloading and mmWaves in D2D Communication. In addition, an approach that solves the joint implementation of all D2D technical challenges is lacking. Furthermore, further research is needed on D2D Intelligent approaches utilizing Multi-hop relay D2D (more than two hops) and all spectrum and transmission modes. Also, focus should be given to approaches that are modular in terms of Radio Access Technologies (RAT) interfaces and can solve the D2D challenges by using DAI control. With DAI, a dynamic and flexible control of the D2D Network can be achieved, with less computation and no hardware changes at BS. Moreover, an intelligent approach handling the ultra-reliable low latency feature in D2D communication is lacking.

2.4 NEED OF AI IN 5G/6G AND BEYOND

It is becoming commonly accepted that Artificial Intelligence (AI) will be one of the crucial driving forces that will shape the future 6G communication networks in designing and optimizing 6G architectures and protocols which will, among others, enable the proliferation of distributed independent autonomous systems [35, 139]. Latest literature in 6G [139–146] specify that in order for the 6G to satisfy connectivity demands of smart networks and satisfy the requirements of near-future services a fully decentralized control with virtual resources [146] is needed. In addition, 6G will bring intelligence from centralized computing facilities to every terminal in the network. Unsupervised learning, combined with inter-user inter-operator knowledge sharing, will promote real-time network decisions through prediction [144]. Also AI, Deep Learning, Machine learning (i.e., DNN (Deep Neural Network), Q-Learning) will help 6G for establishing self-organization strategies, including self-learning, self-configuration, self-healing and self-optimization of network resources at the Terminal level (Mobile Devices), as well as Intelligent Programmable Wireless Environments [147]. Furthermore, distributed security mechanisms will be implemented on mobile devices (i.e., decentralized authentication) and smart mobile applications will be able to learn from user behaviour [141, 145] for improving security and usability. Thus, this research, taking into account the statements above, investigates also the intelligent part of 6G and devises a DAI framework that respects also the implementation of the D2D challenges in future 6G communication networks.

2.5 RELATED WORK

This section provides a review of research work related to the use of BDI agents, other D2D frameworks and transmission mode Selection, and examples of clustering techniques in static and dynamic environments that exist in the open literature related to their usage in telecommunications. Also, related work on the use of unsupervised learning clustering techniques is provided.

2.5.1 RELATED WORK ON UTILIZING BDI AGENTS FOR TELECOMMUNICATIONS PROBLEMS

There is a wealth of research on the use of AI and ML techniques for communication and networking issues. In this section, we include a a few examples that deal with the use of multi-agent systems and BDI agents in general communication problems.

2.5.1.1 Multi-Agent Approaches for Wireless and Mobile Communications

The authors in Ref. [148] tackle the problem of energy consumption and communication latency in wireless sensor networks. More specifically, the authors propose a system with a single Mobile Agent (MA) travelling freely within the network and performing data collection. This behavior improves data delivery to the sink and reduces energy consumption. The specific work utilizes deep neural network for learning, in which the input is the state of the wireless sensor network and the output is the optimal route path. The route planning can be done with the usage of the locations of each node in the environment that acts as input for the intelligent agent. The intelligent agent architecture selected is the actor network and a critic network. The information used is from the whole network, but the decision is taken locally.

Another work that uses reinforcement learning is [149], which deals with the problem of discovering low-level wireless communication schemes between two agents in a fully decentralized system. This is the type of problem considered in the DARPA Spectrum Collaboration Challenge (SC2), which is "the first-of-its-kind collaborative machine-learning competition to overcome scarcity in the radio frequency (RF) spectrum". The proposed method employs policy gradients to learn an ideal bi-directional communication scheme. The approach places two agents against each other and shows that the two actors are able to learn modulation schemes for communication while sharing only a limited information and having no domain-specific knowledge about the task.

2.5.1.2 BDI Agents for Wireless and Mobile Communications

The authors in Ref. [150] utilize a multi-agent software design, dynamic analysis, and decentralized control in order to implement solutions for the complex distributed systems of Wireless Sensor Networks (WSN). The paper's purpose is to create an autonomic system design for distributed nodes in a diverse and changing environment, that interact on top of a wireless communication channel for decentralized problem solving. Due to hardware limitations, the Multi-agent system techniques and especially nodes (agents) are not deliberative (or strong) reasoning systems. The belief, desire, intention (BDI) agent model is used. The paper authors implement two simple WSN test scenarios and show that BDI agents can perform basic WSN functions. In addition, the agents succeed in imitating some recognizable aspects of the system and they are adaptable to different scenarios. In the scenarios, five different agents are discussed. A problem of this approach is that a better method is needed for managing the size of the belief-base used in each agent, as this turns out to expand unboundedly in a case such as flooding.

Another class of wireless networks built dynamically in an ad-hoc network manner with a large mobile user base is found in Vehicular Ad-Hoc Networks (VANETs). The work presented in Ref. [151] tackles the problem of routing in VANETs. Routing in VANETs is critical because of limitations such as unpredictable network topology, frequent disconnections, and varying network densities. The authors in this paper proposed a Multi-agent scheme-based routing scheme that comprises static agent and mobile agents for VANETs (V2V vehicle-to-vehicle communication) where they tackle the challenge of how to route the data with short communication delay, overhead, and the complexity. The proposed algorithm has the following steps: (i) establish a connectivity pattern between the vehicles; (ii) create a set of Beliefs; (iii) develop the Desires; (iv) execute the Intentions.

2.5.2 RELATED WORK ON D2D FRAMEWORKS

This section provides a literature review related to other D2D frameworks that exist in the open literature. The frameworks identified are grouped based on the problem that they tackle and they are briefly compared with the DAI framework proposed in this book. Specifically, other related D2D frameworks appeared in the existing open literature are found in Refs. [152–163]

The frameworks described in Refs.[152, 154, 159] aim to address the caching perspective of D2D communication network. Specifically [152], tries to handle a mobile Content Delivery Network (mCDN) with special mobile devices designated to act as caching servers and they implement caching with the use of Optimum Dual-solution Searching Algorithm (ODSA) that handles content popularity and content policies. Thus, the approach depends on the caching servers for decision control. The framework described in Ref. [154] attempts to use a hypergraph framework that designs the caching-based D2D communication plan by taking into consideration the social ties among users, location, and common interests. For establishing the hypergraph, a trade-off between cellular capacity and D2D capacity must be considered, by using hypergraph-based interference management with the use of BS in a centralized manner. The caching strategy is optimized with the constraints of hit ratio, delay, and caching capacity for improving energy efficiency and spectrum efficiency. The third framework [159] forms a centralized area controller (CAC) that takes content-aware decisions for content access requests in a distributed manner with the use of a Distributed D2D controller (DDC). This is implemented with Q-Learning in a DAI manner; however, it heavily depends on the BS in order to conclude for an estimate.

The framework described in Ref. [153] aims to address the security perspective of a D2D network. Specifically, in the framework, a secure Network-Assisted D2D framework is proposed, which provides a protocol that runs over all the D2D communication network under the BS. The approach achieves security with the creation of a coalition list on specific cases according to states (coalition/non-coalition) and coverage where the BS is taking a major position. This framework always consults the BS for any security-wise decision.

The frameworks described in Refs. [155–158, 160–162] aim to address the generation of D2DSHRs under the BS. Specifically, in Ref. [155] the framework enables the network-assisted scheduling. The framework is not only considering the Base Station to collect D2D and cellular information but also considering the information gathered by any mobile user under the BS. The framework described in Ref. [156], targets the optimal network partition for D2D multicast offloading. The purpose here is to minimize the overall energy consumption at the mobile terminal. In the third framework [157] the authors boost the data rate in D2D communication by enabling data sharing among users with the use of cooperative multicast transmission and with the help of the BS. In the framework described in Ref. [158], the authors propose a D2D opportunistic relay selection greedy algorithm with QoS enforcement that handles offloading and relaying with the use of ProSe. Next, in Ref. [160] the authors realize a 5G cellular system based on D2D communication and four levels of cloud units with various hardware capabilities utilized at the edge of the cellular network as in mobile edge computing (MEC). In Ref. [161] the authors implemented a clustering and topological interference management (TIM) algorithm for a D2D communication network by splitting the mobile network into various groups where each group is served on a different frequency. The authors consider the TIM as a low-rank-matrix-completion problem (LRMC) problem and tackles it using a low-complexity scheme based on semi-definite programming (SDP). Finally, the framework described in Ref. [162] selects active smartphones as relays with the purpose to opportunistically collect heartbeat messages from the adjacent smartphones using D2D communication, hence it is energy efficient.

The framework described in Ref. [163] aims to address the disaster recovery problem in a D2D network. The framework "FINDER" discovers and relinks the isolated Mobile Nodes (MNs) in the disaster zone to minimize the damage on assets and number on human life losses in a case of war. More specifically, the MNs under the damaged Base Station (BS) switch to multi-hop D2D communications mode in a disaster and try to be an active/working Mobile network through a neighboring BS or a Wi-Fi access point. This approach is distributed but it depends on the MNs.

2.5.3 RELATED WORK ON TRANSMISSION MODE SELECTION IN D2D COMMUNICATION CONSIDERING A STATIC ENVIRONMENT

In this section, we review open literature approaches related to Transmission Mode selection in D2D communication, where there is a plethora of articles, as for example [105, 164–176]. Below we refer only to those that are most relevant to the work investigated in this chapter.

A classification based on the type of control (see Section 2.3.4.3) appears below:

- Centralized approaches [105, 164–167, 169, 171–175, 177, 178], where the decision is taken by the BS;
- Semi-distributed approaches [168], where the decision is taken by both the BS and the D2D devices in collaboration;
- Distributed approaches with centralized information [170], where the decision is taken by the D2D devices; however in this case the D2D devices need some information from the BS; and
- Distributed Artificial Intelligence (DAI) approaches, where the decision is taken by each D2D device independently; however in this case they may share information with other D2D devices (this book).

It is evident from above that most works use the Centralized approach and only a few use semi- or fully distributed algorithms. Note that the metrics considered by the previously mentioned approaches for selecting the Transmission mode are shown in Table 2.9. Most of the works use the same metrics (power, SINR, distance).

The approaches described in Refs. [105, 164–166, 168] focus on D2D Transmission Mode Selection but for D2D Direct selection mode only. More specifi-

Table 2.9
Metrics Utilized in D2D Transmission Mode Selection

Metrics	Works Using the Metric
Power or transmission power	[167, 169–171, 175]
Interference	[170, 174]
Resource blocks or sub-channel	[105]
SINR	[105, 164, 165, 168, 169, 172]
Channel Signal Indicator (CSI)	[177]
Distance	[164, 167, 178], this book
Hop count (in multi hop relays)	[177]
Sum rate or type of frequencies	[166, 174]
Battery capacity	[178]
Data forwarding delay	[178]
Link throughput to eNB (BS)	[178]
Weighted data rate	this book

cally, in Ref. [164] the authors use only the quality of the cellular link and interference (SINR) and a simple condition to select the best D2D device to connect. In Ref. [165] the authors are also using the SINR, but with the target to maximize the sum rate by using a gradient method. In Ref. [105] the authors, in addition to SINR, consider Sum Rate as well, by utilizing an evolutionary algorithm. In Ref. [166] the aim is to maximize the average Sum Rate by utilizing an opportunistic subchannel scheduling to solve a stochastic optimization problem. The authors in Ref. [168] use SINR and Lagrangian dual decomposition method in conjunction with a greedy and a column generation based algorithm. With this approach a threshold calculation is first executed at the BS (Lagrange multiplier). Then, the UEs based on the calculated threshold perform a decision independently.

The approaches described in Refs. [167, 169–175] focus on D2D Direct and D2DSHR selection mode only. More specifically, in Ref. [167] the authors use the power usage as a metric and propose a distance-dependent algorithm with power optimization based on the UE position. In Ref. [169], using as utility the power and the SINR, the authors select the best D2DSHR by tackling a mixed integer nonlinear programming problem using both a two-dimensional and a three-dimensional matching. In Ref. [170] the authors choose a D2DSHR by utilizing interference as a metric. In this approach, a distributed method is chosen to coordinate the interference and eliminate improper D2DSHRs by minimizing power. In Ref. [171] the authors formulate the D2DSHR selection problem as a combination optimization many-to-one matching problem. Power is used as a metric in their Power efficient Relay Selection algorithm. In Ref. [172], by using SINR as a metric, a two-stage D2DSHR selection is proposed. In the first stage, the range of the candidate D2DSHR UEs are determined by using a regional division method. In the second stage, the optimal

D2DSHR UE is selected. In Ref. [173], by using distance as a metric, a multi-cell model based on stochastic geometry is proposed. The aim of this model is to evaluate the coverage probability of three location-aware relay selection schemes. In Ref. [174], the authors based on outage probabilities analysis and a sum-capacity comparison provide the criteria of employing Relay communication mode with two hops. The metric used in this analysis is interference that is calculated based on Sum Rate. In Ref. [175], by using power as a metric, an iterative Hungarian method (IHM) is proposed to solve the optimal power allocation problem. This method takes under consideration the channel allocation.

The approaches described in Refs. [177,178] focus on D2D Direct and D2DMHR only. More specifically, in Ref. [177], the authors are using graph theory (Destination Oriented Directed Acyclic Graph (DODAG)) to provide, by means of multi-hop path, the location of D2D nodes in the cluster network topology. Initially, by using as a metric the channel state information (CSI), the BS concludes with the potential D2DMHRs and D2D devices. Then, the hop count metric is utilized as a cumulative cost function to construct the graph. In Ref. [178], the authors propose an Ordinal Potential Game (OPG), with the purpose to select the best link and association between D2D nodes. In this approach, the Transmission Mode Selection is performed as a throughput maximization problem with delay and remaining energy constraints. The metrics used for the selection are the location information, battery capacity, data forwarding delay, and the link throughput associated with it to the eNodeB (BS).

2.5.4 RELATED WORK ON TRANSMISSION MODE SELECTION IN D2D COMMUNICATION CONSIDERING A DYNAMIC ENVIRONMENT

In Section 2.5.3 all approaches are focused on a static environment, an environment without consideration of the mobility of the devices. On the other hand, the DAIS approach can be utilized in dynamic environments, as it is distributed, autonomous, dynamic, flexible, and reacts fast and adapts quickly and efficiently to D2D Network topology changes.

Additionally, to the best of our knowledge, not a lot of work was done in directly addressing Dynamic Transmission Mode Selection. An interesting heuristic algorithmic approach appears in Ref. [179]. It uses only two D2D modes, the: (i) D2D Direct mode; and (ii) D2D Relay mode in a reduced distance of 20m, as well as three modes of operation of the UEs, the: (i) infrastructure mode; (ii) D2D mode; and (iii) D2D Relay mode. We label this approach the "D2D Single Hop Relay Approach (SHRA)". The authors implement two experiments in terms of user mobility. First, they have the UEs static location and then the UEs move within a fixed area. Second, they simulate mobility in both models, the random way-point model and the linear mobility model. The examined approach uses only single-hop D2D Relay communications, and it focuses on the distance for selecting the D2D Relay device. More specifically, the examined approach has two thresholds, the minimum "threshold distance for single-hop D2D communication" (called α) used for establishing D2D Relay assisted communication and the maximum "threshold distance for relay-aided D2D communication" (called γ) used for establishing D2D Direct communication.

Based on the distance (called r) among two D2D Devices that want to communicate, they have the following cases:

- If the distance among two D2D Devices is greater than γ, then they select to connect over the BS.
- If the distance is less than γ and greater than α, then they find a D2D Device that should convert to D2D Relay and both devices should connect between them with the use of the relay device.
- If the distance is less than α, then they connect directly among them using D2D Direct mode.

Note that the SHRA approach connects two D2D Devices that want to communicate using two cases:

- In the first case, the devices select the D2D Direct transmission mode, and then they establish a direct link between each other.
- In the second case, the devices select the D2D Client transmission mode. Then the approach locates and utilizes an existing device that will act as an intermediate node to set its transmission mode to D2D Relay. Subsequently, the two D2D clients connect to the identified D2D Relay by establishing direct links to it.

In contrast, with the DAIS (shown in Section 6.1.5) approach, the D2D Relay forms a cluster in the D2D network towards the BS, and all D2D devices are connected through the BS/Gateway. Finally, the simulation evaluation results in Ref. [179] showed that the D2D Relay mode in the Dynamic and Static environment can provide a better data rate.

2.5.5 RELATED WORK ON UNSUPERVISED LEARNING CLUSTERING TECHNIQUES

In this section, we provide related work on AI/ML Unsupervised Learning Clustering Techniques, utilized for a comparative performance evaluation in our investigation. Since there are no comparative DAI techniques in the open literature addressing D2D Transmission Mode selection, we consider a number of representative AI/ML unsupervised learning clustering techniques, which are parameterized for the D2D environment to allow a fairer comparative evaluation with enhanced DAIS. In particular, we consider Fuzzy ART, DBSCAN, G-MEANS, and MEC clustering techniques. Their performance is evaluated in terms of Spectral Efficiency (SE), Power Consumption (PC), and QoS/QoE metrics. It is important to highlight here that these clustering techniques were not designed for application in D2D communication specifically. With unsupervised learning clustering techniques, an AI classification algorithm, that is associated with generative learning models, may cluster unsorted data according to similarities and differences even if there are no categories provided [180,181]. Below the Fuzzy ART [182–185], DBSCAN [186–189], MEC [190–192] and G-MEANS [181,193,194] clustering techniques, that are implemented and compared with DAIS and DSR, are briefly described. Additionally, the K-means algorithm, with which the Fuzzy ART, MEC and G-Meams approaches are related, is described.

K-Means (Lloyd's algorithm) [195] is a vector quantization method that, by using a set of input patterns, aims to partition n samples (e.g., in our case the number of the UEs in the Network) into K clusters, in which each sample belongs to the cluster with the nearest mean. More specifically, K-Means repeatedly finds the centroid of each cluster in the partition and then re-partitions the input according to which of these centroids is closest. In this setting, the mean operation is an integral over a region of space, and the nearest centroid operation results in clusters. The K-Means is considered as a hard clustering method, in which each sample must be assigned to only one cluster; thus K identifies the coarseness of the partition.

Note that the number of clusters K is a parameter that must be manually set before execution. This is considered as a disadvantage in D2D communication networks which are dynamic in nature. Also, K-Means is slow and with poor results in terms of correct clustering of samples. For these reasons, K-Means is not selected to be examined in the comparative performance evaluation.

Fuzzy ART [183–185] is an unsupervised learning clustering algorithm. It is a type of Adaptive Resonance Theory (ART) network approach [182] which, similarly to K-Means algorithm, uses single prototypes to internally represent and dynamically adjust clusters (as seen in Ref. [195]). However, Fuzzy ART uses as a metric the minimum required similarity between patterns in order to categorize samples in the same cluster. The resulting number of clusters depends on the distances between all input patterns, presented towards the network for the period of training cycles. Fuzzy ART uses structure calculus based on fuzzy logic and ART for binary and continuous value inputs.

DBSCAN [186–189] relies on a density-based concept of clusters which is outlined to determine clusters of uninformed shape. More specifically, for each point of a cluster, the neighborhood of a given radius (called eps; from the greek word "epsilon") has to enclose at least a minimum number of MinPts[9] points. The eps and MinPts are respectively important and mandatory parameters to the algorithm.

In the direction of finding a cluster, it starts with a random point and retrieves all points density-reachable from the chosen point. During the execution of the algorithm, if the selected point is a core point, this procedure results in a cluster. Otherwise, the point is labeled as noise (border). More specifically, if the investigated point contains a sufficient number of points, a cluster is started. The examined point in the algorithm might afterwards be found in a satisfactorily sized radius-environment of a different point and therefore it can be made part of a former cluster. If the selected point is a border point, no points are density-reachable from the selected point and DBSCAN visits the subsequently point. If within the radius of neighborhood the minimum amount of points in the G-neighborhood is not satisfied then the investigated point is considered as non-core point. Precisely, if a point is found to be part of a cluster, its neighborhood is also part of that cluster. Hence, all points that are found within the neighborhood are added, as is their own neighborhood.

The aforesaid process continues until the cluster is found. In that case, a new unvisited point is retrieved and processed, leading to the discovery of a further cluster of noise. Additionally, based on the global values of Eps and MinPts, if two clusters of different density are "close" to each other the DBSCAN algorithm can combine two clusters into one. Accordingly, a recursive call of DBSCAN could be crucial for the identified clusters by means of a higher value on behalf of MinPts. But this is not necessarily a disadvantage for the algorithm because the recursive call of DBSCAN yields a more accurate result of clusters. Moreover, the recursive clustering of the points of a cluster is only crucial under conditions that can be uncomplicatedly recognized with the use of the Euclidean distance.

Minimum Entropy Clustering (MEC) algorithm [190–192], focuses on the minimization of the conditional entropy of clusters, given samples so at the end it concludes with the clusters. Numerous mathematical facts, such as Fano's inequality and Bayes probability of error, indicate that the MEC method can perform well on grouping patterns. This is the reason that MEC: (i) performs well even when the correct number of clusters is unknown; (ii) correctly reveals the structure of data; and (iii) effectively identifies outliers simultaneously. However, MEC is an iterative algorithm starting with an initial partition given by any other (except the random initialization) clustering method (e.g., K-Means), where the number of the clusters formed and the number of clients assigned to each cluster, are values randomly selected. Therefore, in this investigation the initialization is done with the use of the data results coming from the K-Means execution. In addition, the MEC starts with a large K and the algorithm often can automatically remove unnecessary clusters and reach a lower entropy state. This method performs very well especially when the exact number of clusters is unknown. The method can also correctly reveal the structure of data and effectively identify outliers simultaneously with the minimum entropy clustering criterion.

G-MEANS (Gaussian expectation- maximization) clustering [181, 193, 194] extends K-Means approach with the automatic determination of the amount of clusters by normality investigation. The G-MEANS algorithm is based on a statistical experiment for the hypothesis that a subset of data follows a Gaussian distribution. G-MEANS runs K-Means with increasing k hierarchically until the test acknowledges the hypothesis that the data relegated to each K-Means center are Gaussian.

The G-MEANS algorithm begins with a trivial amount of K-Means centers, and steadily grows the amount of centers in each iteration. Specifically, in every iteration of the algorithm, each center whose data do not come from a Gaussian distribution, is separated in two other centers. In spite of the fact that the K-Means algorithm expects, without condition, that the data points in each cluster are spherically distributed around the center, the G-MEANS (Gaussian expectation-maximization) algorithm expects that the data points in each cluster have a multidimensional Gaussian distribution with a covariance matrix (that might or might not be rigid, or mutual). The Gaussian distribution tests are suitable also for covariance matrix assumption. In order to restrict the G-MEANS algorithm from making poor decisions about clusters with few data points, the aforesaid test takes also under consideration the quantity of data points tested by integrating in the calculation the critical value of the test.

The advantages of G-MEANS are that: (i) the hypothesis test does not limit the covariance of the data and (ii) it is not computing a full covariance matrix. The G-MEANS uses the standard statistical significance level of zero.

REFERENCES

1. A. H. Bond and L. Gasser, *Readings in Distributed Artificial Intelligence*. San Francisco, CA: Morgan Kaufmann Publishers Inc., 2014.
2. N. Fornara, "Among autonomous agents in multiagent systems," Ph.D. thesis, University of Lugano, New York, 2003, pp. 1–18, Jun. 2003.
3. S. K. Srivastava, "Applications of intelligent agents," *Electronics Information and Planning*, vol. 26, no. 5, pp. 273–281, 1999.
4. A. M. Andrew, *Foundations of Distributed Artificial Intelligence*, vol. 28. Hoboken, NJ: Wiley, 1999.
5. A. H. Bond, "A survey of distributed artificial intelligence," in *Readings in Distributed Artificial Intelligence* (A. H. Bond and L. Gasser, eds.). Burlington, MA: Morgan Kaufmann, pp. 1–57, 1988.
6. "Distributed Artificial Intelligence — by Francesco Corea — Medium." [Online]. Available at: https://francesco-ai.medium.com/distributed-artificial-intelligence-3e3491e0771c Accessed on: 2021-07-28.
7. E. H. Durfee, "Distributed problem solving and planning," in *EASSS* (M. Luck, V. Marík, O. Stepánková, and R. Trappl, eds.). London: Springer, pp. 118–149, 2001.
8. W. Yeoh and M. Yokoo, "Distributed problem solving," *AI Magazine*, vol. 33, pp. 53–65, 2012.
9. E. H. Durfee, V. R. Lesser, and D. D. Corkill, "Trends in cooperative distributed problem solving," *IEEE Transactions on Knowledge and Data Engineering*, vol. 1, no. 1, pp. 63–83, 1989.
10. G. Beni and J. Wang, "Swarm intelligence in cellular robotic systems," in *Robots and Biological Systems: Towards a New Bionics?* (P. Dario, G. Sandini, and P. Aebischer, eds.). Berlin Heidelberg: Springer, pp. 703–712, 1993.
11. S. Ponomarev and A. E. Voronkov, "Multi-agent systems and decentralized artificial superintelligence," *ArXiv*, vol. abs/1702.08529, feb 2017.
12. R. R. Dumke, S. Mencke, and C. Wille, *Quality Assurance of Agent-Based and Self-Managed Systems*. Boca Raton, FL: CRC Press, Jan. 2009.
13. H. V. D. Parunak, "Applications of distributed artificial intelligence in industry," *Foundations of Distributed Artificial Intelligence*, vol. 116, no. 92, pp. 139–164, 1994.
14. P. Stone and M. Veloso, "Multiagent systems: A survey from a machine learning perspective," *Autonomous Robots*, vol. 8, pp. 345–383, Jun. 2000.
15. G. Weiss, *Multiagent Systems: A Modern Approach to Distributed Artificial Intelligence*, vol. 3. Gerhard Weiss, Cambridge: MIT Press, 1999.
16. Y. Demazeau, and J.-P. Muller, *Decentralized A.I.* North-Holland, Elsevier, 1990.
17. M. Herlihy and N. Shavit, *The Art of Multiprocessor Programming*. Burlington, MA: Morgan Kaufmann, 2012.
18. I. Foster, *Designing and Building Parallel Programs: Concepts and Tools for Parallel Software Engineering*. Boston, MA: Addison-Wesley, 1995.
19. M. Mikolajczak, "Designing and building parallel programs: Concepts and tools for parallel software engineering," *IEEE Concurrency*, vol. 5, no. 2, pp. 88–90, 2005.

20. G. O'Hare and N. Jennings, *Foundations of Distributed Artificial Intelligence*. Hoboken, NJ: Wiley, 1996.
21. A. S. Rao and M. P. Georgeff, "BDI agents: From theory to practice.," *ICMAS*, vol. 95, pp. 312–319, 1995.
22. J. R. Stuart and N. Peter, *Artificial Intelligence a Modern Approach*, vol. 72. Cambridge: Cambridge University Press (CUP), 2003.
23. Y. Shoham and K. Leyton-Brown, *Multiagent Systems: Algorithmic, Game-Theoretic, and logical Foundations*. Cambridge, MA: Cambridge University Press, 2008.
24. P. Caillou, B. Gaudou, A. Grignard, C. Q. Truong, and P. Taillandier, "A simple-to-use BDI architecture for agent-based modeling and simulation," in *Advances in Social Simulation 2015* (W. Jager, R. Verbrugge, A. Flache, G. de Roo, L. Hoogduin, and C. Hemelrijk, eds.). Cham: Springer International Publishing, pp. 15–28, 2017.
25. E. Amouroux, T. Q. Chu, A. Boucher, and A. Drogoul, "GAMA: An environment for implementing and running spatially explicit multi-agent simulations," in *Lecture Notes in Computer Science (Including Subseries Lecture Notes in Artificial Intelligence and Lecture Notes in Bioinformatics)*, vol. 5044. Berlin, Heidelberg: Springer, pp. 359–371, 2009.
26. S. Alam, B. Edmonds, and R. Meyer, *Agent Computing and Multi-Agent Systems*, vol. 5044 of *Lecture Notes in Computer Science*. Berlin, Heidelberg: Springer, 2009.
27. Y. Shoham and K. Leyton-Brown, *Multiagent Systems*. Cambridge: Cambridge University Press (CUP), 2008.
28. "Welcome to the Foundation for Intelligent Physical Agents." [Online]. Available at: http://www.fipa.org/ Accessed on: 2021-02-27.
29. C. Paper and R. H. Bordini, *BDI Agent Programming in AgentSpeak Using Jason*. Berlin Heidelberg: Springer, 2006.
30. K. Doppler, M. P. Rinne, P. änis, C. Ribeiro, and K. Hugl, "Device-to-device communications; Functional prospects for LTE-advanced networks," *Proceedings - 2009 IEEE International Conference on Communications Workshops, ICC 2009*, Dresden, Jun. 2009.
31. G. Fodor, E. Dahlman, G. Mildh, S. Parkvall, N. Reider, G. Miklós, and Z. Turányi, "Design aspects of network assisted device-to-device communications," *IEEE Communications Magazine*, vol. 50, pp. 170–177, Mar. 2012.
32. J. Liu, Y. Kawamoto, H. Nishiyama, N. Kato, and N. Kadowaki, "Device-to-device communications achieve efficient load balancing in LTE-advanced networks," *IEEE Wireless Communications*, vol. 21, no. 2, pp. 57–65, 2014.
33. G. Fodor and N. Reider, "A distributed power control scheme for cellular network assisted D2D communications," in *2011 IEEE Global Telecommunications Conference - GLOBECOM 2011*, pp. 1–6, IEEE, Houston, TX, Dec. 2011.
34. K. Doppler, M. Rinne, C. Wijting, C. B. Ribeiro, and K. Hug, "Device-to-device communication as an underlay to LTE-advanced networks," *IEEE Communications Magazine*, vol. 47, no. 12, pp. 42–49, 2009.
35. N. Rajatheva, I. Atzeni, E. Bjornson, A. Bourdoux, S. Buzzi, J.-B. Doré, S. Erkucuk, M. Fuentes, K. Guan, Y. Hu, X. Huang, J. Hulkkonen, J. Jornet, M. Katz, R. Nilsson, E. Panayirci, K. M. Rabie, N. Rajapaksha, M. Salehi, H. Sarieddeen, T. Svensson, O. Tervo, A. Tolli, Q. Wu, and W. Xu, "White paper on broadband connectivity in 6G," *arXiv: Signal Processing*, vol. abs/2004.14247, 2020.
36. I. F. Akyildiz, S. Nie, S. C. Lin, and M. Chandrasekaran, "5G roadmap: 10 key enabling technologies," *Computer Networks*, vol. 106, pp. 17–48, 2016.

37. M. Noura and R. Nordin, "A survey on interference management for device-to-device (D2D) communication and its challenges in 5G networks," *Journal of Network and Computer Applications*, vol. 71, pp. 130–150, Aug. 2016.
38. A. Asadi, V. Mancuso, and P. Jacko, "Floating band D2D: Exploring and exploiting the potentials of adaptive D2D-enabled networks," *Proceedings of the WoWMoM 2015: A World of Wireless Mobile and Multimedia Networks*, Boston, MA, Jun. 2015.
39. J. Deng, A. A. Dowhuszko, R. Freij, and O. Tirkkonen, "Relay selection and resource allocation for D2D-relaying under uplink cellular power control," *2015 IEEE Globecom Workshops, GC Wkshps 2015 - Proceedings*, San Diego, CA, Dec. 2015.
40. G. Steri, G. Baldini, I. N. Fovino, R. Neisse, and L. Goratti, "A novel multi-hop secure LTE-D2D communication protocol for IoT scenarios," *2016 23rd International Conference on Telecommunications, ICT 2016*, Thessaloniki, May 2016.
41. L. Song, D. Niyato, Z. Han, and E. Hossain, "Game-theoretic resource allocation methods for device-to-device communication," *IEEE Wireless Communications*, vol. 21, no. 3, pp. 136–144, 2014.
42. B. Peng, T. Peng, Z. Liu, Y. Yang, and C. Hu, "Cluster-based multicast transmission for device-to-device (D2D) communication," *IEEE Vehicular Technology Conference*, Las Vegas, NV, Sep. 2013.
43. H. Claussen, "Performance of macro- and co-channel femtocells in a hierarchical cell structure," in *2007 IEEE 18th International Symposium on Personal, Indoor and Mobile Radio Communications*, pp. 1–5, Athens, Sep. 2007.
44. H. B. Valiveti and P. T. Rao, "EHSD: An exemplary handover scheme during D2D communication based on decentralization of SDN," *Wireless Personal Communications*, vol. 94, no. 4, pp. 2393–2416, 2017.
45. F. Wang, C. Xu, L. Song, and Z. Han, "Energy-efficient resource allocation for device-to-device underlay communication," *IEEE Transactions on Wireless Communications*, vol. 14, no. 4, pp. 2082–2092, 2015.
46. J. Yue, C. Ma, H. Yu, and W. Zhou, "Secrecy-based access control for device-to-device communication underlaying cellular networks," *IEEE Communications Letters*, vol. 17, pp. 2068–2071, Nov. 2013.
47. C. Ma, W. Wu, Y. Cui, and X. Wang, "On the performance of successive interference cancellation in D2D-enabled cellular networks," in *2015 IEEE Conference on Computer Communications (INFOCOM)*, pp. 37–45, IEEE, Hong Kong, Apr. 2015.
48. K. Zia, N. Javed, M. N. Sial, S. Ahmed, H. Iram, and A. A. Pirzada, "A survey of conventional and artificial intelligence/learning based resource allocation and interference mitigation schemes in D2D enabled networks," *ArXiv*, vol. abs/1809.08748, 2018.
49. E. Hossain and M. Hasan, "5G cellular: Key enabling technologies and research challenges," *IEEE Instrumentation & Measurement Magazine*, vol. 18, pp. 11–21, Jun. 2015.
50. X. Ge, S. Tu, G. Mao, C.-X. Wang, and T. Han, "5G ultra-dense cellular networks," *IEEE Wireless Communications*, vol. 23, pp. 72–79, Feb. 2016.
51. J. Liu, S. Zhang, N. Kato, H. Ujikawa, and K. Suzuki, "Device-to-device communications for enhancing quality of experience in software defined multi-tier LTE-A networks," *IEEE Network*, vol. 29, pp. 46–52, Aug. 2015.
52. M. Sadik, N. Akkari, and G. Aldabbagh, "SDN-based handover scheme for multi-tier LTE/femto and D2D networks," *Computer Networks*, vol. 142, pp. 142–153, 2018.
53. Y. Niu, Y. Li, D. Jin, L. Su, and A. V. Vasilakos, "A survey of millimeter wave communications (mmWave) for 5G: Opportunities and challenges," *Wireless Networks*, vol. 21, pp. 2657–2676, Nov. 2015.

54. C. Liaskos, S. Nie, A. Tsioliaridou, A. Pitsillides, S. Ioannidis, and I. Akyildiz, "A new wireless communication paradigm through software-controlled metasurfaces," *IEEE Communications Magazine*, vol. 56, no. 9, pp. 162–169, 2018.
55. P. Gandotra and R. Jha, "Device-to-device communication in cellular networks: A survey," *Journal of Network and Computer Applications*, vol. 71, no. 4, pp. 1801–1819, 2016.
56. P. Gandotra, R. Kumar Jha, and S. Jain, "A survey on device-to-device (D2D) communication: Architecture and security issues," *Journal of Network and Computer Applications*, vol. 78, pp. 9–29, Jan. 2017.
57. F. Jameel, Z. Hamid, F. Jabeen, S. Zeadally, and M. A. Javed, "A survey of device-to-device communications: Research issues and challenges," *IEEE Communications Surveys and Tutorials*, vol. 20, no. 3, pp. 2133–2168, 2018.
58. J. Liu, N. Kato, J. Ma, and N. Kadowaki, "Device-to-device communication in LTE-advanced networks: A survey," *IEEE Communications Surveys & Tutorials*, vol. 17, no. 4, pp. 1923–1940, 2015.
59. B. Jedari, F. Xia, and Z. Ning, "A survey on human-centric communications in non-cooperative wireless relay networks," *IEEE Communications Surveys and Tutorials*, vol. 20, no. 2, pp. 914–944, 2018.
60. R. I. Ansari, C. Chrysostomou, S. A. Hassan, M. Guizani, S. Mumtaz, J. Rodriguez, and J. J. P. C. Rodrigues, "5G D2D networks: Techniques, challenges, and future prospects," *IEEE Systems Journal*, vol. 12, pp. 3970–3984, Dec. 2018.
61. S. Ullah, T. LeAnh, and C. Hong, "On delay minimization of layered video streaming in icn enabled cellular networks with D2D communication," in *On Delay Minimization of Layered Video Streaming in ICN Enabled Cellular Networks with D2D Communication*, vol. 8, pp. 1157–1159, ICN, Dec. 2017.
62. M. Bagaa, A. Ksentini, T. Taleb, R. Jantti, A. Chelli, and I. Balasingham, "An efficient D2D-based strategies for machine type communications in 5G mobile systems," *IEEE Wireless Communications and Networking Conference, WCNC*, pp. 1–6, Doha, Qatar, Sep. 2016.
63. X. Zhang and Q. Zhu, "Statistical QoS provisioning over D2D-offloading based 5G multimedia big-data mobile wireless networks," *INFOCOM 2018 - IEEE Conference on Computer Communications Workshops*, pp. 742–747, Honolulu, HI, Apr. 2018.
64. X. Wang, Y. Zhang, V. C. Leung, N. Guizani, and T. Jiang, "D2D big data: Content deliveries over wireless device-to-device sharing in large-scale mobile networks," *IEEE Wireless Communications*, vol. 25, no. 1, pp. 32–38, 2018.
65. Y. He, F. R. Yu, N. Zhao, and H. Yin, "Secure social networks in 5G systems with mobile edge computing, caching, and device-to-device communications," *IEEE Wireless Communications*, vol. 25, no. 3, pp. 103–109, 2018.
66. S. Kazemi Rashed, R. Shahbazian, and S. A. Ghorashi, "Learning-based resource allocation in D2D communications with QoS and fairness considerations," *Transactions on Emerging Telecommunications Technologies*, vol. 29, no. 1, pp. 1–20, 2018.
67. A. Moussaid, W. Jaafar, W. Ajib, and H. Elbiaze, "Deep reinforcement learning-based data transmission for D2D communications," *2018 14th International Conference on Wireless and Mobile Computing, Networking and Communications (WiMob)*, pp. 1–7, Limassol, Nov. 2018.
68. K. Zia, N. Javed, M. N. Sial, S. Ahmed, A. A. Pirzada, and F. Pervez, "A distributed multi-agent RL-based autonomous spectrum allocation scheme in D2D enabled multi-tier HetNets," *IEEE Access*, vol. 7, pp. 6733–6745, 2019.

69. S. H. Chen, A. J. Jakeman, and J. P. Norton, "Artificial intelligence techniques: An introduction to their use for modelling environmental systems," *Mathematics and Computers in Simulation*, vol. 78, no. 2–3, pp. 379–400, 2008.
70. M.-Y. Cheng, H.-C. Tsai, and C.-L. Liu, "Artificial intelligence approaches to achieve strategic control over project cash flows," *Automation in Construction*, vol. 18, pp. 386–393, Jul. 2009.
71. H. v. Hasselt, A. Guez, and D. Silver, "Deep reinforcement learning with double q-learning," in *Proceedings of the Thirtieth AAAI Conference on Artificial Intelligence, AAAI'16*, pp. 2094—2100, AAAI Press, Phoenix, AZ, 2016.
72. C. J. C. H. Watkins and P. Dayan, "Q-learning," *Machine Learning*, vol. 8, pp. 279–292, May 1992.
73. Y. Bengio, "Learning deep architectures for AI," *Foundations and Trends in Machine Learning*, vol. 2, pp. 1–127, Nov. 2009.
74. D. J. Russo, B. Van Roy, A. Kazerouni, I. Osband, and Z. Wen, "A tutorial on thompson sampling," *Foundations and Trends in Machine Learning*, vol. 11, pp. 1–96, Jul. 2018.
75. D. Russo and B. V. Roy, "An information-theoretic analysis of thompson sampling," *Journal of Machine Learning Research*, vol. 17, no. 1, pp. 2442–2471, 2016.
76. "Thompson Sampling & Bayesian Control Rule [Pedro A. Ortega]." [Online]. Available at: http://www.adaptiveagents.org/bayesian˙control˙rule Accessed on: 2019-05-03.
77. G. Winter, J. Periaux, M. Galan, and P. Cuesta, *Genetic Algorithms in Engineering and Computer Science*. New York: Wiley, 1995.
78. T. Back, *Evolutionary Algorithms in Theory and Practice: Evolution Strategies, Evolutionary Programming, Genetic Algorithms*. Oxford: Oxford University, 1996.
79. A. P. Engelbrecht, *Computational Intelligence: An Introduction*. Hoboke, NJ: John Wiley & Sons, 2007.
80. A. Y. Zomaya, *Handbook of Nature-Inspired and Innovative Computing: Integrating Classical Models with Emerging Technologies*. Berlin, Heidelberg: Springer, 2006.
81. M. A. Rahman, Y. Lee, and I. Koo, "Energy-efficient power allocation and relay selection schemes for relay-assisted D2D communications in 5G wireless networks," *Sensors (Switzerland)*, vol. 18, no. 9, 2018.
82. V. Novák, I. Perfilieva, and J. Mockor, "Fuzzy logic: What, why, for which?," in *Mathematical Principles of Fuzzy Logic* (V. Novák, I. Perfilieva, and J. Mockor, eds.). Boston, MA: Springer US, pp. 1–14, 1999.
83. Y. Luo, P. Hong, and R. Li, "Energy efficiency-delay tradeoff in energy-harvesting-based D2D communication: An experimental learning approach," *IEEE Communications Letters*, vol. 22, no. 8, pp. 1704–1707, 2018.
84. S. Toumi, M. Hamdi, and M. Zaied, "An adaptive q-learning approach to power control for D2D communications," in *2018 International Conference on Advanced Systems and Electric Technologies, IC˙ASET 2018*, pp. 206–209, Hammamet, Mar. 2018.
85. Y.-F. Huang, T.-H. Tan, N.-C. Wang, Y.-L. Chen, and Y.-L. Li, "Resource allocation for D2D communications with a novel distributed Q-learning algorithm in heterogeneous networks," *2018 International Conference on Machine Learning and Cybernetics (ICMLC)*, vol. 2, pp. 533–537, Chengdu, Jul. 2018.
86. Y. Qiu, Z. Ji, Y. Zhu, G. Meng, and G. Xie, "Joint mode selection and power adaptation for D2D communication with reinforcement learning," *2018 15th International Symposium on Wireless Communication Systems (ISWCS)*, pp. 1–6, Lisbon, Aug. 2018.

87. Y.-F. Huang, T.-H. Tan, and Y.-L. Li, "Performance of resource allocation for D2D communications in Q-learning based heterogeneous networks," *2018 IEEE International Conference on Consumer Electronics-Taiwan (ICCE-TW)*, pp. 1–5, Taichung, May 2018.
88. Y. Luo, Z. Feng, H. Jiang, Y. Yang, Y. Huang, and J. Yao, "Game-theoretic learning approaches for secure D2D communications against full-duplex active eavesdropperr," *IEEE Access*, vol. 7, pp. 41324–41335, 2019.
89. Y. Luo, M. Zeng, and H. Jiang, "Learning to tradeoff between energy efficiency and delay in energy harvesting-powered D2D communication: A distributed experience-sharing algorithm," *IEEE Internet of Things Journal*, vol. 6, no. 3, pp. 5585–5594, 2019.
90. "An Introduction to Q-Learning: Reinforcement Learning." [Online]. Available at: https://medium.freecodecamp.org/an-introduction-to-q-learning-reinforcement-learning-14ac0b4493cc, Accessed on: 2018-10-17.
91. "Artificial Intelligence: Foundations of Computational Agents – 11.3.3 Q-Learning.' [Online]. Available at: https://artint.info/html1e/ArtInt˙265.html
92. J. Xu, X. Gu, and Z. Fan, "D2D power control based on hierarchical extreme learning machine," *2018 IEEE 29th Annual International Symposium on Personal, Indoor and Mobile Radio Communications (PIMRC)*, pp. 1–7, Bologna, Sep. 2018.
93. W. Lee, M. Kim, and D.-H. Cho, "Transmit power control using deep neural network for underlay device-to-device communication," *IEEE Wireless Communications Letters*, vol. 8, pp. 141–144, Feb. 2019.
94. P. A. Ortega and D. A. Braun, "Generalized thompson sampling for sequential decision-making and causal inference," *Complex Adaptive Systems Modeling*, vol. 2, no. 1, pp. 1–25.
95. K. I. Ahmed, H. Tabassum, and E. Hossain, "Deep learning for radio resource allocation in multi-cell networks," *IEEE Network*, vol. 33, pp. 188–195, 2019.
96. W. Lee, M. Kim, and D. H. Cho, "Deep learning based transmit power control in underlaid device-to-device communication," *IEEE Systems Journal*, vol. 13, pp. 2551–2554, Sep. 2019.
97. Z. Li and C. Guo, "A multi-agent deep reinforcement learning based spectrum allocation framework for D2D underlay communications," *arXiv: Networking and Internet Architecture*, vol. abs/1912.09302, 2019.
98. A. Abdelreheem, O. A. Omer, H. Esmaiel, and U. S. Mohamed, "Deep learning-based relay selection in D2D millimeter wave communications," in *2019 International Conference on Computer and Information Sciences, ICCIS 2019*, pp. 1–5, Aljouf, Apr. 2019.
99. M. Nielsen, "Neural networks and deep learning," *Machine Learning*, 2015. Available: http://neuralnetworksanddeeplearning.com, Accessed on: 2024-08-29.
100. I. Goodfellow, Y. Bengio, and A. Courville, "*Introduction*." in *Deep Learning*. Cambridge, MA: The MIT Press, pp. 1–12, 2015.
101. S. J. Darak, H. Zhang, J. Palicot, and C. Moy, "An efficient policy for D2D communications and energy harvesting in cognitive radios: Go Bayesian!," *2015 23rd European Signal Processing Conference, EUSIPCO 2015*, pp. 1231–1235, Nice, Aug. 2015.
102. P. Viappiani, A. Gopalan Aditya, S. Mannor Shie, and Y. Mansour Mansour, "Thompson sampling for complex online problems," *ICML*, pp. 399–410, Nov. 2014.
103. P. A. Ortega, D. A. Braun, and S. Godsill, "Reinforcement learning and the Bayesian control rule," *AGI'11: Proceedings of the 4th International Conference on Artificial General Intelligence*, Springer-Verlag, vol. 6830, pp. 281–285, Mountain View, CA, 2011.

104. P. Ortega, K.-E. Kim, and D. Lee, "Reactive bandits with attitude," *Journal of Machine Learning Research*, vol. 38, pp. 1–14, 2015.
105. H. Pang, P. Wang, X. Wang, F. Liu, and N. N. Van, "Joint mode selection and resource allocation using evolutionary algorithm for device-to-device communication underlaying cellular networks," *Journal of Communications*, vol. 8, no. 11, pp. 751–757, 2013.
106. P. A. Vikhar, "Evolutionary algorithms: A critical review and its future prospects," *Proceedings of the International Conference on Global Trends in Signal Processing, Information Computing and Communication, ICGTSPICC 2016*, pp. 261–265, Jalgaon, Dec. 2017.
107. D. Corne and M. A. Lones, "Evolutionary algorithms," in *Handbook of Heuristics* (R. C. Martí, P. M. Pardalos, and M. G. C. Resende, eds.). Berlin, Heidelberg: Springer, pp. 409–430, 2018.
108. M. Hamdi, D. Yuan, and M. Zaied, "GA-based scheme for fair joint channel allocation and power control for underlaying D2D multicast communications," *2017 13th International Wireless Communications and Mobile Computing Conference, IWCMC 2017*, pp. 446–451, Jun. 2017.
109. C. Yang, X. Xu, J. Han, and X. Tao, "GA based user matching with optimal power allocation in D2D underlaying network," *IEEE Vehicular Technology Conference*, pp. 1–5, Jan. 2014.
110. D. Whitley, "A genetic algorithm tutorial," *Statistics and Computing*, vol. 4, no. 2, pp. 65–85, 1994.
111. J. Shen, O. Wang, and J. Yu, "Non-orthogonal pulse shape modulation for wireless relay D2D communication," *Proceedings of the 2017 International Conference on Computing Intelligence and Information System, CIIS 2017*, pp. 86–92, Jan. 2018.
112. Z. Huang, Z. Zeng, H. Xia, and J. Shi, "Power control in two-tier OFDMA femtocell networks with particle swarm optimization," *IEEE Vehicular Technology Conference*, pp. 1–5, May 2011.
113. L. Su, Y. Ji, P. Wang, and F. Liu, "Resource allocation using particle swarm optimization for D2D communication underlay of cellular networks," *IEEE Wireless Communications and Networking Conference, WCNC*, pp. 129–133, Apr. 2013.
114. S. Sun and Y. Shin, "Resource allocation for D2D communication using particle swarm optimization in LTE networks," *International Conference on ICT Convergence*, pp. 371–376, 2014.
115. R. Tang, J. Dong, Z. Zhu, J. Liu, J. Zhao, and H. Qu, "Resource allocation for underlaid device-to-device communication by incorporating both channel assignment and power control," *Proceedings of the 2015 5th International Conference on Communication Systems and Network Technologies, CSNT 2015*, pp. 432–436, Apr. 2015.
116. Y. F. Huang, T. H. Tan, S. H. Liu, and Y. F. Chen, "Performance of resource allocation in device-to-device communication systems based on particle swarm optimization," *2017 IEEE International Conference on Systems, Man, and Cybernetics, SMC 2017*, pp. 400–404, Jan. 2017.
117. T. Liang, T. Zhang, J. Cao, and C. Feng, "Joint resource allocation and power control scheme for device-to-device communication underlaying cellular networks," *2014 International Symposium on Wireless Personal Multimedia Communications (WPMC)*, pp. 568–572, Sep. 2014.
118. J. Xu, C. Guo, and J. Yang, "Bio-inspired power control and channel allocation for cellular networks with D2D communications," *Wireless Networks*, vol. 25, no. 3, pp. 1273–1288, 2019.

119. J. Kennedy and R. Eberhart, "Particle swarm optimization," *Proceedings of ICNN'95: International Conference on Neural Networks*, pp. 1942–1948, Apr. 1995.
120. P. Wilson, H. A. Mantooth, P. Wilson, and H. A. Mantooth, "Model-based optimization techniques," in *Model-Based Engineering for Complex Electronic Systems* (P. Wilson and H. A. Mantooth, eds.). Oxford: Newnes, pp. 347–367, Jan. 2013.
121. R. Eberhart and J. Kennedy, "A new optimizer using particle swarm theory," in *MHS'95. Proceedings of the Sixth International Symposium on Micro Machine and Human Science*, pp. 39–43, Oct. 1995.
122. M. Tanha, D. Sajjadi, F. Tong, and J. Pan, "Disaster management and response for modern cellular networks using flow-based multi-hop device-to-device communications," *IEEE Vehicular Technology Conference*, pp. 1–7, Sep. 2017.
123. S. Feki, A. Masmoudi, A. Belghith, F. Zarai, and M. S. Obaidat, "Swarm intelligence-based radio resource management for D2D-based V2V communication," *International Journal of Communication Systems*, pp. 1–16, May 2018.
124. T. S. Marco Dorigo, *Ant Colony Optimization*. Main St Bradford, PA: A Bradford Book, 2004.
125. A. Shekhawat, P. Poddar, and D. Boswal, "Ant colony optimization algorithms: Introduction and beyond," *Indian Institute of Technology Bombay: Artificial Intelligence Seminar 2009*, p. 41, 2009.
126. M. Dorigo, M. Birattari, and T. Stutzle, "Ant colony optimization," *IEEE Computational Intelligence Magazine*, vol. 1, no. 4, pp. 28–39, 2006.
127. D. Karaboga and B. Basturk, "A powerful and efficient algorithm for numerical function optimization: Artificial bee colony (ABC) algorithm," *Journal of Global Optimization*, vol. 39, no. 3, pp. 459–471, 2007.
128. M. Dorigo and T. Stützle, *Ant Colony Optimization*, Cambridge: MIT Press, 2004.
129. M. R. Jalali, A. Afshar, and M. A. Marino, "Ant colony optimization algorithm (ACO); A new heuristic approach for engineering optimization," *WSEAS Transactions on Information Science and Applications*, vol. 2, no. 5, pp. 606–610, 2005.
130. M. Condoluci, L. Militano, A. Orsino, J. Alonso-Zarate, and G. Araniti, "Lte-direct vs. wifi-direct for machine-type communications over LTE-A systems," in *2015 IEEE 26th Annual International Symposium on Personal, Indoor, and Mobile Radio Communications (PIMRC)*, pp. 2298–2302, Sep. 2015.
131. T. Koskela, S. Hakola, T. Chen, and J. Lehtomäki, "Clustering concept using device-to-device communication in cellular system," *IEEE Wireless Communications and Networking Conference, WCNC*, pp. 0–5, Apr. 2010.
132. S. Models, "Stochastic Models 10.1," in *Stochastic Models*, p. 723, North-Holland, 1971.
133. A. Elgabli, H. Khan, M. Krouka, and M. Bennis, "Reinforcement learning based scheduling algorithm for optimizing age of information in ultra reliable low latency networks," *2019 IEEE Symposium on Computers and Communications (ISCC)*, pp. 1–6, Jun. 2019.
134. O. Gonzalez, H. Shrikumar, J. A. Stankovic, and K. Ramamritham, "Adaptive fault tolerance and graceful degradation under dynamic hard real-time scheduling," *Proceedings - Real-Time Systems Symposium*, pp. 79–89, Dec. 1997.
135. L. Jorguseski, A. Pais, F. Gunnarsson, A. Centonza, and C. Willcock, "Self-organizing networks in 3GPP: Standardization and future trends," *IEEE Communications Magazine*, vol. 52, no. 12, pp. 28–34, 2014.

136. J. Moysen and L. Giupponi, "From 4G to 5G: Self-organized network management meets machine learning," *Computer Communications*, vol. 129, pp. 248–268, 2018.
137. O. Bello, S. Zeadally, and S. Member, "Intelligent device-to-device communication in the internet of things," *IEEE Xplore*, vol. 10, pp. 1–11, Sep. 2015.
138. L. P. Kaelbling, M. L. Littman, and A. W. Moore, "Reinforcement learning: A survey," *Journal of Artificial Intelligence Research*, vol. 4, pp. 237–285, Apr. 1996.
139. R.-A. Stoica and G. Abreu, "6G: The wireless communications network for collaborative and AI applications," *ArXiv*, vol. abs/1904.03413, 2019.
140. K. B. Letaief, W. Chen, Y. Shi, J. Zhang, and Y.-J. A. Zhang, "The roadmap to 6G: AI empowered wireless networks," *IEEE Communications Magazine*, vol. 57, no. 8, pp. 84–90, 2019.
141. I. F. Akyildiz, A. Kak, and S. Nie, "6G and beyond: The future of wireless communications systems," *IEEE Access*, vol. 8, pp. 133995–134030, 2020.
142. R.-L. Aguiar, "White paper for research beyond 5G," Technical Report, Oct. 2015.
143. K. David and H. Berndt, "6G vision and requirements: Is there any need for beyond 5G?," *IEEE Vehicular Technology Magazine*, vol. 13, no. 3, pp. 72–80, 2018.
144. M. Giordani, M. Polese, M. Mezzavilla, S. Rangan, and M. Zorzi, "Toward 6G networks: Use cases and technologies," *IEEE Communications Magazine*, vol. 58, no. 3, pp. 55–61, 2020.
145. E. C. Strinati, S. Barbarossa, J. González-Jiménez, D. Kténas, N. Cassiau, and C. Dehos, "6G: The next frontier," *ArXiv*, vol. abs/1901.03239, 2019.
146. S. Yrjola, "Decentralized 6G business models," in *6G Wireless Summit*. Finland: IEEE, pp. 1–2, 2019.
147. C. Liaskos, A. Tsioliaridou, S. Nie, A. Pitsillides, S. Ioannidis, and I. Akyildiz, "An interpretable neural network for configuring programmable wireless environments," in *2019 IEEE 20th International Workshop on Signal Processing Advances in Wireless Communications (SPAWC)*, pp. 1–5, Jul. 2019.
148. J. Lu, L. Feng, J. Yang, M. M. Hassan, A. Alelaiwi, and I. Humar, "Artificial agent: The fusion of artificial intelligence and a mobile agent for energy-efficient traffic control in wireless sensor networks," *Future Generation Computer Systems*, vol. 95, pp. 45–51, 2019.
149. C. de Vrieze, S. Barratt, D. Tsai, and A. Sahai, "Cooperative multi-agent reinforcement learning for low-level wireless communication," *ArXiv*, vol. abs/1801.04541, pp. 1–15, 2018.
150. A. Morris, P. Giorgini, and S. Abdel-Naby, "Simulating BDI-based wireless sensor networks," in *2009 IEEE/WIC/ACM International Joint Conference on Web Intelligence and Intelligent Agent Technology*, vol. 2, pp. 78–81, Sep. 2009.
151. M. S. Kakkasageri, M. J. Sataraddi, P. M. Chanal, and G. S. Kori, "BDI agent based routing scheme in vanets," in *2017 International Conference on Wireless Communications, Signal Processing and Networking (WiSPNET)*, pp. 129–133, Mar. 2017.
152. H. J. Kang and C. G. Kang, "Mobile device-to-device (D2D) content delivery networking: A design and optimization framework," *Journal of Communications and Networks*, vol. 16, no. 5, pp. 568–577, 2014.
153. A. Ometov, P. Masek, J. Urama, J. Hosek, S. Andreev, and Y. Koucheryavy, "Implementing secure network-assisted D2D framework in live 3GPP LTE deployment," *2016 IEEE International Conference on Communications Workshops, ICC 2016*, pp. 749–754, May 2016.

154. B. Bai, L. Wang, Z. Han, W. Chen, and T. Svensson, "Caching based socially-aware D2D communications in wireless content delivery networks: A hypergraph framework," *IEEE Wireless Communications*, vol. 23, no. 4, pp. 74–81, 2016.
155. L. Pu, X. Chen, J. Xu, and X. Fu, "D2D fogging: An energy-efficient and incentive-aware task offloading framework via network-assisted D2D collaboration," *IEEE Journal on Selected Areas in Communications*, vol. 34, no. 12, pp. 3887–39014, 2016.
156. S. Yu, R. Langar, and X. Wang, "A D2D-multicast based computation offloading framework for interactive applications," *2016 IEEE Global Communications Conference, GLOBECOM 2016 - Proceedings*, pp. 16–21, Dec. 2016.
157. S. Doumiati, H. Artail, and K. Kabalan, "A framework for clustering LTE devices for implementing group D2D communication and multicast capability," *2017 8th International Conference on Information and Communication Systems, ICICS 2017*, pp. 216–221, Apr. 2017.
158. A. Asadi, V. Mancuso, and R. Gupta, "Dore: An experimental framework to enable outband D2D relay in cellular networks," *arXiv*, vol. 25, no. 5, pp. 2930–2943, 2017.
159. N. Kumar, S. N. Swain, and C. Siva Ram Murthy, "A novel distributed Q-learning based resource reservation framework for facilitating D2D content access requests in LTE-A networks," *IEEE Transactions on Network and Service Management*, vol. 15, no. 2, pp. 718–731, 2018.
160. A. A. Ateya, A. Muthanna, and A. Koucheryavy, "5G framework based on multi-level edge computing with D2D enabled communication," *International Conference on Advanced Communication Technology, ICACT*, vol. 2018, pp. 507–512, Feb. 2018.
161. S. Doumiati, M. Assaad, and H. A. Artail, "A framework of topological interference management and clustering for D2D networks," *IEEE Transactions on Communications*, vol. 67, pp. 7856–7871, Nov. 2019.
162. X. Yi, L. Pan, Y. Jin, F. Liu, and M. Chen, "EDirect: Energy-efficient D2D-assisted relaying framework for cellular signaling reduction," *IEEE/ACM Transactions on Networking*, vol. 28, no. 2, pp. 860–873, 2020.
163. A. Thomas and G. Raja, "FINDER: A D2D based critical communications framework for disaster management in 5G," *Peer-to-Peer Networking and Applications*, vol. 12, no. 4, pp. 912–923, 2019.
164. K. Doppler, C. H. Yu, C. B. Ribeiro, and P. Jänis, "Mode selection for device-to-device communication underlaying an LTE-advanced network," *IEEE Wireless Communications and Networking Conference, WCNC*, Apr. 2010.
165. M. Jung, K. Hwang, and S. Choi, "Joint mode selection and power allocation scheme for power-efficient device-to-device (D2D) communication," *IEEE Vehicular Technology Conference*, pp. 1–5, May 2012.
166. M. H. Han, B. G. Kim, and J. W. Lee, "Subchannel and transmission mode scheduling for D2D communication in OFDMA networks," *IEEE Vehicular Technology Conference*, pp. 1–5, Sep. 2012.
167. S. Xiang, T. Peng, Z. Liu, and W. Wang, "A distance-dependent mode selection algorithm in heterogeneous D2D and IMT-advanced network," *2012 IEEE Globecom Workshops, GC Wkshps 2012*, pp. 416–420, Dec. 2012.
168. Y. Liu, "Optimal mode selection in D2D-enabled multibase station systems," *IEEE Communications Letters*, vol. 20, no. 3, pp. 470–473, 2016.
169. C. Xu, J. Feng, B. Huang, Z. Zhou, S. Mumtaz, and J. Rodriguez, "Joint relay selection and resource allocation for energy-efficient D2D cooperative communications using matching theory," *Applied Sciences (Switzerland)*, vol. 7, no. 5, pp. 1–24, 2017.

170. X. Ma, R. Yin, G. Yu, and Z. Zhang, "A distributed relay selection method for relay assisted device-to-device communication system," *IEEE International Symposium on Personal, Indoor and Mobile Radio Communications, PIMRC*, pp. 1020–1024, Sep. 2012.
171. B. Ma, H. Shah-Mansouri, and V. W. Wong, "A matching approach for power efficient relay selection in full duplex D2D networks," *2016 IEEE International Conference on Communications, ICC 2016*, pp. 1–6, May 2016.
172. M. Zhao, X. Gu, D. Wu, and L. Ren, "A two-stages relay selection and resource allocation joint method for D2D communication system," *IEEE Wireless Communications and Networking Conference, WCNC*, pp. 1–6, Sep. 2016.
173. H. Feng, H. Wang, X. Chu, and X. Xu, "On the tradeoff between optimal relay selection and protocol design in hybrid D2D networks," *2015 IEEE International Conference on Communication Workshop, ICCW 2015*, pp. 705–711, Jun. 2015.
174. L. Wang, T. Peng, Y. Yang, and W. Wang, "Interference constrained D2D communication with relay underlaying cellular networks," *IEEE Vehicular Technology Conference*, pp. 1–5, Sep. 2013.
175. T. Kim and M. Dong, "An iterative Hungarian method to joint relay selection and resource allocation for D2D communications," *IEEE Wireless Communications Letters*, vol. 3, no. 6, pp. 625–628, 2014.
176. S. M. Kazmi, N. H. Tran, W. Saad, Z. Han, T. M. Ho, T. Z. Oo, and C. S. Hong, "Mode selection and resource allocation in device-to-device communications: A matching game approach," *IEEE Transactions on Mobile Computing*, vol. 16, no. 11, pp. 3126–3141, 2017.
177. G. Rigazzi, F. Chiti, R. Fantacci, and C. Carlini, "Multi-hop D2D networking and resource management scheme for M2M communications over LTE-A systems," *IWCMC 2014 - 10th International Wireless Communications and Mobile Computing Conference*, pp. 973–978, Aug. 2014.
178. J. Gui and J. Deng, "Multi-hop relay-aided underlay D2D communications for improving cellular coverage quality," *IEEE Access*, vol. 6, pp. 14318–14338, 2018.
179. U. N. Kar and D. K. Sanyal, "Experimental analysis of device-to-device communication," *2019 12th International Conference on Contemporary Computing, IC3 2019*, pp. 1–6, Aug. 2019.
180. D. Wang and G. Hinton, "Review of unsupervised learning," *AI Magazine*, vol. 22, no. 2, pp. 101–102, 2001.
181. N. Li, M. Shepperd, and Y. Guo, "A systematic review of unsupervised learning techniques for software defect prediction," *Information and Software Technology*, vol. 122, p. 106287, Feb. 2020.
182. T. Frank, K. Kraiss, and T. Kuhlen, "Comparative analysis of fuzzy art and ART-2A network clustering performance," *IEEE Transactions on Neural Networks*, vol. 9, pp. 544–559, May 1998.
183. G. Aydin Keskin, S. Ilhan, and C. Özkan, "The Fuzzy ART algorithm: A categorization method for supplier evaluation and selection," *Expert Systems with Applications*, vol. 37, pp. 1235–1240, Mar. 2010.
184. G. A. Carpenter, S. Grossberg, and D. B. Rosen, "Fuzzy ART: Fast stable learning and categorization of analog patterns by an adaptive resonance system," *Neural Networks*, vol. 4, pp. 759–771, Jan. 1991.
185. S. G. Akojwar and R. M. Patrikar, "Real time classifier for industrial wireless sensor network using neural networks with wavelet preprocessors," in *Proceedings of the IEEE International Conference on Industrial Technology*, pp. 512–517, Dec. 2006.

186. M. Ester, H.-P. Kriegel, J. Sander, and X. Xu, "A density-based algorithm for discovering clusters in large spatial databases with noise," in *KDD'96: Proceedings of the Second International Conference on Knowledge Discovery and Data Mining*, Portland, Oregon, pp. 226–231, AAAI Press, 1996.
187. M. Li, D. Meng, S. Gu, and S. Liu, "Research and improvement of DBSCAN cluster algorithm," *Proceedings of the 2015 7th International Conference on Information Technology in Medicine and Education, ITME 2015*, pp. 537–540, Nov. 2016.
188. C. Dharni and M. Bnasal, "An improvement of DBSCAN algorithm to analyze cluster for large datasets," *Proceedings of the 2013 IEEE International Conference in MOOC, Innovation and Technology in Education, MITE 2013*, pp. 42–46, Dec. 2013.
189. K. Khan, S. U. Rehman, K. Aziz, S. Fong, S. Sarasvady, and A. Vishwa, "DBSCAN: Past, present and future," *5th International Conference on the Applications of Digital Information and Web Technologies, ICADIWT 2014*, pp. 232–238, Feb. 2014.
190. S. J. Roberts, C. Holmes, and D. Denison, "Minimum-entropy data clustering using reversible jump Markov chain Monte Carlo," *Lecture Notes in Computer Science (Including Subseries Lecture Notes in Artificial Intelligence and Lecture Notes in Bioinformatics)*, vol. 2130, no. 8, pp. 103–110, 2001.
191. H. Li, K. Zhang, and T. Jiang, "Minimum entropy clustering and applications to gene expression analysis," *Proceedings of the 2004 IEEE Computational Systems Bioinformatics Conference, 2004 (CSB 2004)*, pp. 142–151, Aug. 2004.
192. F. Golchin and K. K. Paliwal, "Minimum-entropy clustering and its application to lossless image coding," *IEEE International Conference on Image Processing*, vol. 2, pp. 262–265, Oct. 1997.
193. G. Hamerly and C. Elkan, "Learning the k in k-means," in *Proceedings of the 16th International Conference on Neural Information Processing Systems, NIPS'03*, Cambridge, MA, pp. 281–288, MIT Press, 2003.
194. A. Smiti and Z. Elouedi, "WCOID-DG: An approach for case base maintenance based on weighting, clustering, outliers, internal detection and Dbsan-Gmeans," *Journal of Computer and System Sciences*, vol. 80, no. 1, pp. 27–38, 2014.
195. M. J. Zaki and M. J. Meira, *Data Mining and Analysis: Fundamental Concepts and Algorithms*. Cambridge, MA: Cambridge University Press, 2013.

Notes

[1] In parallel computing, a perfectly parallel workload is the case where little or no manipulation is needed to separate the examined problem into a number of parallel tasks [17]. This is frequently the case where there is little or no dependency, or need for communication among those parallel tasks [18–20].

[2] Higher persistence to continue current actions independently and with lower persistence to be adaptable and reactive but with inconsistent and computational costly behaviors.

[3] The modeler is a way to define the BDI agents properties through a BDI programming framework. An example of Modeler is the JADE.

[4] This is called also Dynamic Implementation at features.

[5] In parallel computing, a perfectly parallel workload can be considered the case where little or no manipulation is needed to separate the problem into a number of parallel tasks [17]. This is often the case where there is little or no dependency or need for communication between those parallel tasks, or for results between them [18–20].

[6] With the implementation of DAI, agents are independent without any restrictions for forming immediate D2D networks. With DAI the problem is separated and distributed to the nodes in the network. Then each node, in synchronization with others, tries to solve the small portion of the problem assigned to it. At the end all nodes provide the solution they found to the small problem assigned to them, which in aggregate form the solution of the biggest problem.

[7]Self-organizing network (SON) [135, 136] manages networks with high automation that automatically tune the network parameters to improve the network Key Performance Indicators (KPIs). There are three categories of SON: Self-configuration, Self-optimization and Self-healing. Self-healing, the ability to automatically recover from failures, includes detection, diagnosis, and recovery.
[8]Reinforcement learning (RL) [138] is an area of ML concerned with how software agents ought to take actions in an environment so as to maximize some notion of cumulative reward. Reinforcement learning is one of three basic ML paradigms, alongside supervised learning and unsupervised learning.
[9]MinPts are the minimum number of points in the G-neighborhood of a core point.

3 A Novel DAI Framework and BDI Extended Agents

This chapter introduces the proposed BDIx-based DAI framework to tackle 5G/6G challenges in mobile communication networks.[1] It also extends the BDI agents to BDIx agents to allow flexibility in the design of the Plan Library and the realization of dynamic decisions with the use of fuzzy logic IF-THEN statements along with the use of reinforcement learning. Furthermore, it discusses how the framework is distributed,[2] and provides the main features of the BDIx agents and their architecture. Additionally, the realization and implementation aspects of the BDIx agents in the DAI framework according to specific mobile communication network requirements are also discussed. Moreover, the DAI framework implementation, requirements, and characteristics are also discussed. Finally, it provides the operation complexity of the DAI framework.

3.1 THE BDIX-BASED DISTRIBUTED AI FRAMEWORK

In this book, we consider a 5G/6G mobile communication network setting and motivate the implementation of a distributed, autonomous, dynamic, and flexible distributed artificial intelligent (DAI) framework that utilizes BDIx agents (with reinforcement learning), where a BDIx agent resides on each UE (see Figure 3.1).

The proposed BDIx-based DAI framework is expected to offer a number of attractive features, including: (i) fast network control with less messaging exchange, hence

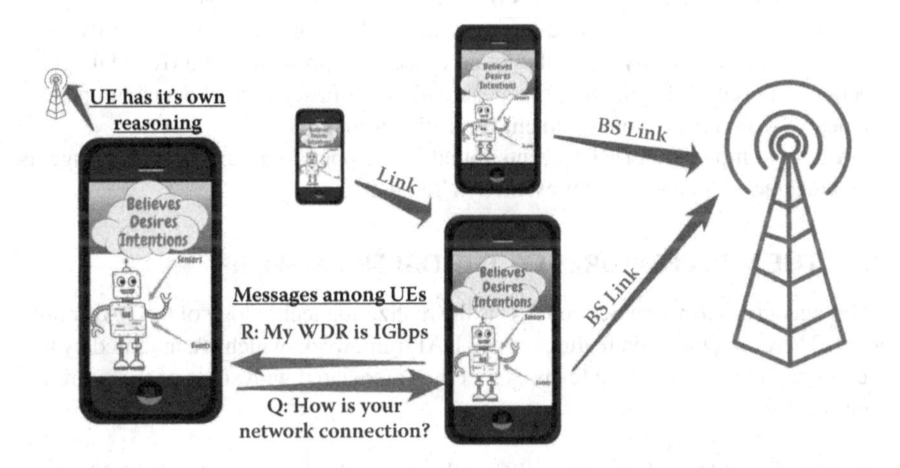

Figure 3.1 The DAI framework: BDIx agents residing on the mobile devices.

DOI: 10.1201/9781003469209-3

a reduced signalling overhead; (ii) fast decision-making; (iii) support of self-healing mechanisms and to collaboratively act as a self-organizing network; and (iv) to capitalize on existing implementations (e.g., artificial neural networks [1]) for tackling any mobile networking challenge. In order to achieve these features, the framework's architecture is envisioned to be modular and utilize the DAI concept. The underlying attraction is that this framework can act as a glue platform in employing any one or more of the optimized intelligent approaches found in the literature, relying only on local knowledge (e.g., use deep neural networks to identify best frequency that reduces interference to be used by an entering D2D device). Thus, targeted modules within the BDIx agents can be substituted or added as (extra AI/ML models) to achieve a specific task/requirement in 5G communication (e.g., to achieve high data rate in a D2D setup). Also, with the use of the BDIx agents in the framework, intercommunication and collaborative decisions can be achieved with the use of messages. It is worth noting that there are a lot of predefined well-structured languages for BDI agents communication, including propose, notify, and inform.

Next, we present a detailed description of the DAI framework and its implementation using BDIx agents. The DAI framework is analyzed and described in more depth, elaborating on: (i) what a BDIx agent is, and how the DAI framework is realized with the use of BDIx agents; (ii) the DAI framework features, as inherited by the use of BDIx agents; (iii) the DAI architecture and flow of operation; and (iv) how the DAI framework can accomplish the mobile communication network's challenges with the use of BDI agents.

Before elaborating on the DAI framework, we provide the main reasons for selecting BDIx agents (a technology first introduced in the 1980s [2]) to realize our proposed DAI framework: (i) The current technology specifications of a CPU (processing power), memory (cheap and plentiful) and networking equipment of a mobile device can be compared with the technology specification of a regular desktop. Thus, BDIx agents can nowadays run easily on a market-based mobile device [3, 4]; (ii) AI/ML, which also characterizes BDIx agents, is improved vigorously within the latest years, and it is widely used in research; (iii) BDI agents can successfully communicate asynchronously and collaborate in tackling problems; and (iv) BDI agents operate with reduced signalling overhead and much faster control decision updates, as they rely on the local environment for decision-making.

A brief comparison of BDI agents found in the open literature, and BDIx agents implemented in this book are presented in Table 3.1.

3.2 THE MAIN FEATURES OF THE DAI FRAMEWORK

BDIx agent (described in Section 3.4) is the realization technology of the DAI framework. Therefore, the main features of our DAI framework, which are inherited by this technology along with some features specifics associated with some of them, are the following:

Modularity: The BDIx agent allows the networks operators to: (i) Add or remove Desires at run time through specific APIs; (ii) Change the relations between

Table 3.1
Summary of BDI and BDIx Agents Differences

Features	BDI Agent	BDIx Agent
Utilizes other AI/ML approaches at beliefs	N	Y
Uses Fuzzy logic with priorities values on beliefs	N	Y
Filters sensor values and raised events	N	Y
Provides REST API to Telcom operators	N	Y
Has LEGO based components	N	Y
Provides concurrent execution of multiple intentions	N	Y
Provides ACID mechanism for beliefs	N	Y
Has an architecture for the implementation	Simpler architecture	Y
Has a flowchart of execution that support the above	Simpler flowchart	Y
Enforces through the BDIx interpreter the whole implementation of the DAI framework	No supported yet	Y
Provides additional features based on the 5G/6G requirements	Specific features	Y
Adapts the characteristics to be aligned with the requirements	N	Y

Beliefs and Desires (through threshold values) that results in the selection of Intentions and the execution of plans (as shown in Ref. [5]).

Multitasking Execution: Multiple problems can be solved concurrently by the BDIx agent with the parallel execution of multiple Intentions. This feature can provide the ability to the proposed framework to achieve a joint implementation of the Challenges (more details appear in Section 4.4).

Collaborative Environment: The BDIx agents can communicate among them using well-defined standard Agent Communication Languages (FIPA ACL/Angel Speak).[3] Consequently, through communication[4] the agents can coordinate and form a collaborative environment through which:

> The BDIx agents can negotiate the acceptance of a proposal by other agents and commit to do their proposed task by considering their Beliefs and Desires. For simplicity, but without loss of generality, in our investigation, we consider that BDIx agents accept the proposals of other agents without considering their own Beliefs and Desires.
>
> Moreover, the LTE proximity services messages from UE devices are not encrypted and are shared freely among the UE devices in the network.

Logging of User Actions: The BDIx agents can gather the actions (tractability) of the UEs owners in terms of bandwidth usage and time in a log table under their Beliefs. Then, agents can use this information to improve the QoE of the user by adjusting priorities of Desires though the Plan Library. With this feature, the agents can also keep history of actions in Beliefs.

Autonomicity:[5] The BDIx agent that is installed in each UE, decides for the control of communication without any dependency on information other than the local information provided by Device Discovery (Proximity Services). Thus, the BDIx agent is responsible for controlling the user's device and network connection.

Dynamicity:[6] The BDIx agent supports reinforcement learning (as shown in Ref. [21]) with the use of sensors and metrics that measure the environment and updates the Beliefs according to the representation of the environment. Additionally, the agent decisions depend only on information it can access as a device through the use of protocols (i.e., proximity services). This feature provides the ability to the proposed framework to handle situations like disaster recovery or emergencies (i.e., ambulance video transmission where the video has pre-specified needs of a specific bandwidth and time delay).

Flexibility:[7] With the use of APIs (REST, Simple Object Access Protocol called SOAP), the framework allows an operator to change the agents Desires and Plan Library "on the fly" (as shown in Section 3.6.5). Initially, the BDIx agents that reside on UEs have some pre-specified Beliefs, Desires and Planning Library's fuzzy logic rule set for setting priorities of Desires. These can be initialized based on the operator's objectives (e.g., to achieve 5G D2D communication) during the process of the device registration in its network. Also, the aforesaid settings can be changed dynamically by the operator for the alignment of the agent with the current objectives of the telecom operator. Therefore, BDIx agents can have updates regularly based on telecom preferences.

Supports Distributed Artificial Intelligence (DAI) Control: The communication control is executed, in a distributed way using local environmental information, by the BDIx agent running on the device. With DAI control, we can break the investigated problem into smaller pieces/requirements (that do not depend on other agents' decision) and achieve 5G and beyond communication collaboratively. Additionally, with the aforementioned segmentation, in which a piece is represented by a Desire and a plan associated with it, the complexity of the communication is reduced.

Supports Security: Each BDIx agent can utilize well-known security techniques, as for example Rivest–Shamir–Adleman (RSA) encryption or Secure Sockets Layer (SSL) protocol along with digital signatures assigned in each device as tools to increase security. This can be exploited for the implementation of a security protocol that will achieve secure communication. In order to further improve security, the communication encryption can be enhanced with the use of Public Key cryptography, Sim Data and Digital Signature (in the same manner as [6,7]).

Provide Good Environmental Representation: The BDix agents can achieve an accurate representation of the surrounding environment in the Beliefs with the use of sensors, variables, simple data structures and with the utilization of high complicated data structures (i.e., Neural Networks). Additionally, the BDIx agents can interact with any of the seven layers of International organization of Standardization–Open System Interconnection (ISO-OSI) for acquiring extra network knowledge and improve its environmental representation.

Light Execution: The BDIx agent uses reduced CPU and memory resources for executing tasks. This allows BDIx agents to run efficiently on today's market based smartphones and Internet of Things (IoT) hardware [3,4].
Deliberation: The BDI agents can have an increasing freedom for selecting Desires to become Intentions [8]. With BDIx agents, this deliberation still exists; however, it is slightly restricted by the fuzzy logic rules of the Plan Library of the agent. More details appear next (Section 3.6.1).
Furthermore,as D2D communication is concerned, the DAI framework provides the following:
Supports both Inband and Outband D2D Communication: The BDIx agent is autonomous, dynamic, flexible and more specifically modular. Therefore, it can utilize any available interface and frequency band, either inband or outband, provided by the operator. For example, an agent can use concurrently both a cellular (i.e., Long-Term Evolution LTE) and a Wi-Fi interface, the one for link sharing and the other for connecting towards the Gateway.
Supports all D2D Transmission Modes: The BDIx agent can support all transmission modes (i.e., D2D Relay, D2D Multi Hop Relay, D2D Client, D2D Direct) with the use of LTE Direct (for Inband D2D) and Wi-Fi Direct (for Outband D2D). More precisely, the agent on the device can share its link and act as D2DSHR, D2DMHR or utilize a shared link as D2D Client. Additionally, the agent can connect to the BS and share its link to other D2D devices as D2DMHR device.
The BDIx agents (as self-learning) can learn from the existing D2D-Relay[8] nodes that share information in the D2D network with the use of latest technologies (i.e., LTE proximity services). This can achieve a wider expansion of the environment coverage and the improvement of the data in the Beliefs.
The BDIx agent can easily provide support to Heterogeneous network (HETNET). Because it can utilize, according to cases, its Wi-Fi interface for sharing or connecting to a Wi-Fi Gateway, the same applies to the mobile interface it can use to connect to any type of mobile network (if supported).

3.3 DECENTRALIZATION OF THE DAI FRAMEWORK

In DAI and more specifically in agent's theory, there are various stages of decentralization [9–13]:

Centralized Communication & Centralized Control (CC&CC): Every device talks to a centralized entity (e.g., the base station). Then the centralized entity decides the details in terms of control and who talks to whom in terms of communication (e.g., Transmission Mode Selection in D2D).
Decentralized Communication & Centralized Control (DC&CC): Devices talk to each other and/or to a centralized entity. A centrally decided algorithm that resides at the centralized entity, decides who talks to whom (Dominating Set Agents).

Decentralized Communication & Decentralized Control (DC&DC): Devices talk to each other and/or to a centralized entity. Each device has control on where to connect (Multi Agent Systems).

The proposed DAI framework utilizes the Decentralized Communication & Decentralized Control with the use of collaborative agents that accept any proposed actions. Note, that we adopt the term distributed. instead of decentralized, as it better conveys the implemented nature of the proposed DAI framework.

3.4 INTRODUCTION TO BDIX AGENTS

A BDIx agent is a BDI agent (see Section 2.1.2) that is extended to utilize in Beliefs any other AI/ML techniques (e.g., fuzzy logic, deep learning neural networks, as shown in Ref. [14]) that gives, among others, a better understanding of the surrounding environment to the agent, as well as the ability to prioritize the order of execution of the Desires (see Figure 3.2).

Specifically, as some Desires must conclude before the execution of others (i.e., because the output of one Desire can be an input to another), we allow the Desires to be assigned with priority values, ranging from 0 (lowest) to 100 (highest) (as shown in Ref. [16]). Furthermore, a Plan Library (with the use of priority values in Desires) must also be used for controlling the execution of Desires and thus restrict agent deliberation so that Intentions can change at run time.

In our framework, this priority value is estimated with the use of a Plan Library [9] implemented with fuzzy logic considering in its "IF-THEN" rules the current

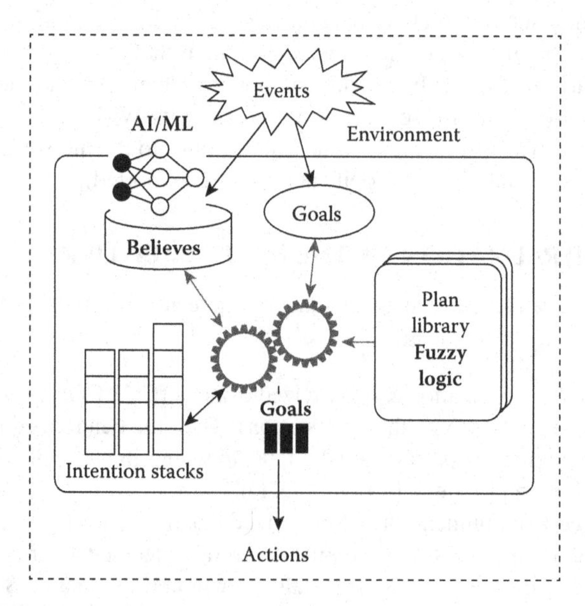

Figure 3.2 BDIx agent with Fuzzy logic & machine learning at beliefs. (Adapted from Ref. [15].)

Beliefs, the values measured by the sensors, and any raised events and cases where the pre-specified threshold values (e.g., the data rate drops to less than 60% in a D2D device) are exceeded [5, 17, 18]. Based on the assigned priority value, Desires become Intentions which are adopted for active pursuit by the agent (referred to as a Goal).

In addition, a Desire that will become an Intention can have multiple plans associated with it and the Desire can select an appropriate plan based on a utility function. For simplicity, but without loss of generality, in our DAI framework, we consider each Desire, and indirectly each Intention, to be associated with only one plan. Therefore, BDIx agents can have an agent environment consisting of Beliefs, Desires, Intentions and plans with a direct relation among them. The sensors can change the BDIx agent's Belief values and raise events. An event may update Beliefs' values, Desires' priorities, trigger plans of Intentions or modify goals (i.e., Intentions that are currently executing).

Note that Beliefs and Desires of a BDIx agent can be changed/extended, at any time and on the fly, according to future needs of the network operator or future changes affecting the network structure or policies. This is a flexibility offered by the proposed DAI framework. It is also worth mentioning that the fuzzy logic residing in the Belief part, acts as a perception part of a BDIx agent. For example, in case of a raised event, fuzzy logic considers the Beliefs of the BDIx agent to select appropriate Desire(s), increasing their related priority and thus becoming Intention(s). It is also important to highlight here that for a less abstract illustration the Beliefs and the Desires of the BDIx agent have been extracted from the D2D requirements/challenges in D2D communication for 5G and beyond, and appear in Chapter 4.

3.4.1 REINFORCEMENT LEARNING IN BDIX AGENTS

Furthermore, our DAI framework supports reinforcement learning (RL), that is, learning what to do and how to map situations to actions so as to maximize a numerical reward [19], by selecting at the BDI agent an appropriate Desire to become Intention and execute a specific Plan (as shown in Ref. [20]). For example, a BDI agent can be enhanced with an Adaptive Neuro Fuzzy Inference system (ANFIS) realized with "Knowledge Acquisition module" (KAM) and RL, as shown in Ref. [21], targeting the improvement of the reactive, proactive and intelligent behaviors in complex applications. In their implementation, the execution of the agent plans is based on the weighted learning by interaction and changes in the beliefs, where the BDI agent interacts with the environment in terms of observing events and learning whether to proceed with the committed intention or look for any other alternatives.

Likewise, in the BDIx agent RL implementation, the agent perceives the resulting changes of actions in terms of data rate, targeting the achievement of QoS and QoE with the use of Back-Propagation Neural Network in the Believes. The implementation details for D2D appear in Section 4.7. In our implementation, RL focuses on QoS and QoE, as these are critical factors in the successful implementation of a telecommunication network.

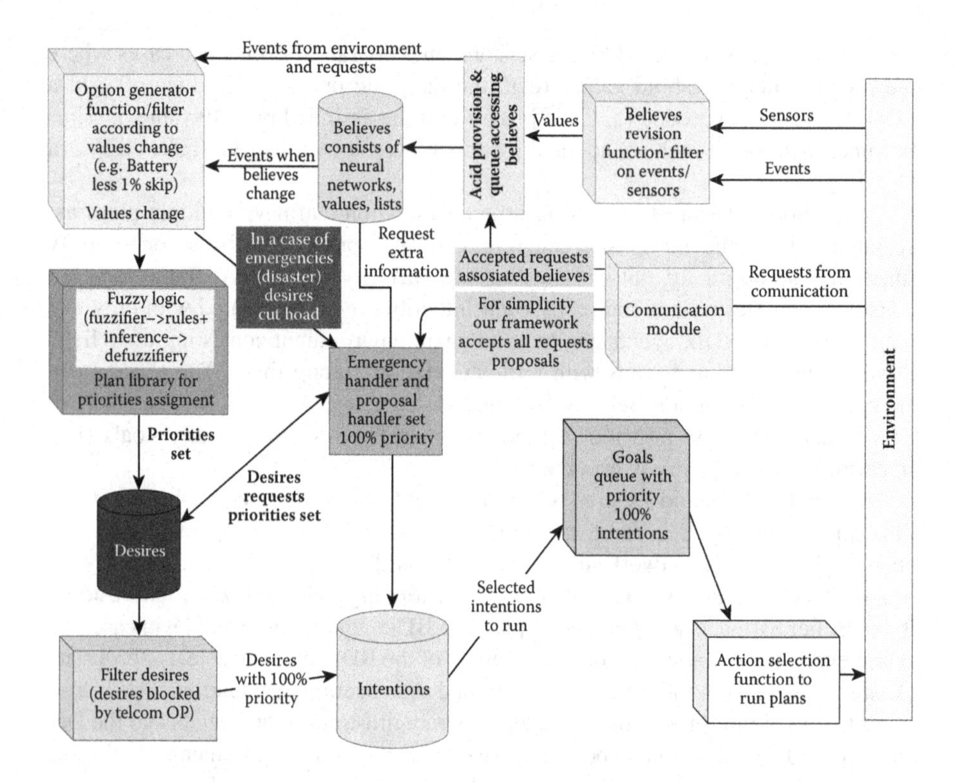

Figure 3.3 BDIx architecture.

3.5 BDIX AGENT ARCHITECTURE

The architectural process model of the BDIx agent is shown in Figure 3.3, with arrows representing data flow from the sensors, the events raised, and the messages from other BDIx agents. It is worth pointing out that the BDIx interpreter runs and accesses the whole BDIx agent architecture.

The model consists of the following components, that are identified by their color, dimension, and shape:

1. The cylinders at the model represent lists such as Beliefs, Desires, and Intentions.
2. The rose color Rectangle is for the "Emergency Handler and Proposal Handler" that will bypass the fuzzy logic rules and it will immediate convert a Desire to Intention with 100% priority. This is a component that exists for immediate case handling such as incidence response to a disaster (e.g., BS stop working) or requests from other BDIx agents. This component has its own procedures according to specific incident or agent request, and also it has direct access for informative purposes to Beliefs.
3. The "shade of gray" blue (cornflower blue) three-dimension rectangles are the filtering process handled by the BDIx interpreter, that filter the inputs from sen-

sors, events and communication among the BDIx agents that are represented with arrows (as described in Sections 3.6.1 and 3.6.4).

4. The olive color orthogonal rectangle monitors data flow that have as their target to change Beliefs, and it is responsible for the ACID operation at the Beliefs (as described in the Section 3.6.2).
5. The blue color three-dimension rectangle represents the Plan Library with the Fuzzifier, Rules & Inference and DeFuzzifier.
6. The "light" brown three-dimension rectangle is a component that implements a filter that filters a specific set of Desires. This filter is defined by the Operator in order to restrict specific undesired Desires by the operator to become Intentions at the specific time.
7. The grey three-dimension rectangle is a component where the Intention Plans are executed (as described in Section 3.6.3).

3.6 REALIZATION AND IMPLEMENTATION ASPECTS OF THE BDIX AGENTS

In this section, we elaborate on a number of implementation specific aspects of BDIx agents (refer to Figure 3.3). The aim of this section is to provide a better understanding of the internal workings of the BDIx agent in terms of a realization and implementation that targets 5G and beyond communication. These include the Deliberation & Persistence of the BDIx agents, the Plan Library and Intentions Concurrent Execution, the flowchart of Intention Execution by Plan Library, and the BDIx Interpreter. Additionally, this section shows how telecom operators can control the BDIx agents.

3.6.1 DELIBERATION AND PERSISTENCE OF THE BDIX AGENTS

The deliberation of a BDIx agent is determined by the implementation target of the DAI framework (e.g. the realization of the D2D Challenges in this book), which have a restrict order of execution in order to achieve a goal. Thus, the objective is to restrict the BDIx agent to make as Intentions only those Desires that are associated with specific communication challenges. With our DAI framework, this is achieved in a deterministic manner by the fuzzy logic rules of the Plan Library of the agent (see Section 3.4), which controls the order of the Desire's execution by setting priority values in Desires. The fuzzy logic is selected because it has the ability by using natural language to capture the expertise of the network operators, it can handle imprecision and hence can robustify the system response. Approaches that use priority values, do not use AI/ML for calculating the prioritization rather use a function of the current step, which would make it more and more probable to be selected based on Beliefs when the execution of the agent progresses [16]. Other approaches uses AI/ML for plan selection connected to goals with the use of fusion ART following a different architecture that what we propose [14].

Additionally, the BDI "persistence" characteristic can affect the BDIx agents' performance and moreover the DAI framework. The reason is that mobile networks are very dynamic in nature affecting the validity of the agents' decisions. Therefore, in our approach, the agents are less persistent, allowing the Intentions to change in real time and be executed according to a Plan Library. Moreover, in order for the framework to tackle the problematic and computationally expensive behaviour caused by the reduced agent' persistence (e.g., infinite loop in selecting transmission mode, instant connections, disconnections and re-connections, etc.), a filtering algorithm can be used for filtering out the unnecessary feedback that comes from the sensors that affect the Beliefs using threshold values. For simplification of the DAI framework, at the startup of each Intention, the Intention gets a persistence value of M (assigned by a constant value of 10^9) that represents the persistence coefficient (as shown in Section 2.1.3). Additionally, the utility function of persistence coefficients is a function that decreases the persistence value by one (1) in each execution of the task. If the persistence value reaches zero, then the intention is marked with priority zero and it becomes a Desire.

3.6.2 PLAN LIBRARY AND INTENTIONS CONCURRENT EXECUTION

The purpose of the Plan Library, with the use of priority values, is to: (i) restrict the deliberation in Agent aiming to keep it "light" in execution and (ii) pre-specify and restrict the order of execution of Intentions (Desires and indirectly plans) with the use of fuzzy logic, aiming to direct the order of execution to the achievement of the 5G communication.

In order to keep BDIx agents "light" in terms of resources, and hence run on mobile devices, the agent is restricted to concurrently pursuit up to a maximum number N of executions of Intentions at the same time. For example, in our implementations, we allow up to 10^{10} Intentions to be concurrently executed by the agent. To achieve this, a Goals Queue (Data Structure) is utilized to keep the N currently running Intentions. Also, the Intentions Queue is used in order for the Plan Library to handle the excess of the Desires with 100% priority values. The purpose is not to restrict the Plan Library (i.e., fuzzy logic) to assign priority values of 100% only to N Desires.

To restrict the order of Intentions execution, priority values on the associated Desires are used. The agent selects a Desire to become an Intention, only if its priority value is equal to 100%. Note that up to N concurrent Intentions can be under the active pursuit of the agent and when finalized, their related Desire's priority value is set to 0%. Then, the Plan Library using fuzzy logic, selects the Desires that should be executed next and increases their priority value accordingly. Also, it is worth noting that some Desires might need to be always treated as Intentions and under the active pursuit of the agent in the DAI framework. This, for example, includes security monitoring and the power reservation of D2D devices (see Chapter 5).

Additionally, due to the concurrent execution of Intentions and the raising of the events, the Beliefs can be changed during run-time, resulting in data inconsistency. To avoid this, the use of well-known Locking mechanisms on the Beliefs is a requirement. The locking mechanisms must work in the same way as in database transactions in order to assure the Atomicity, Consistency, Isolation and Durability (ACID)[11] [22].

3.6.3 FLOWCHART OF INTENTION EXECUTION BY PLAN LIBRARY

The flowchart in Figure 3.4 shows the operation of a BDIx agent from the point it receives a message from the environment, until it selects and executes a plan. After perceiving a change in its environment, the agent checks if the Intention must be satisfied or must be changed. If the Intention is not changed, then it continues with the execution of the Intention plan. If the execution is not successful the agent retries again for a maximum number of M attempts (see Section 3.6.1). After that, if the Intention is still not finalized, the agent selects another Intention from the Intentions queue and executes a Plan that is associated with it. In case the queue is empty, the agent increases the priorities of the existing Desires until some of them reach the value of 100% and are then selected by the agent to become Intentions. It is worth pointing out that a Desire that will become an Intention can have multiple plans associated with it. If this is the case then the Desire can select an appropriate plan based on a utility function. For simplicity, but without loss of generality, in our DAI framework, we consider each Desire, and indirectly each Intention, to be associated with only one plan. Also it is worth noting that the same flow of execution is run at the BDIx agent concurrently N times (set at 10; see Section 3.6.2).

3.6.4 BDIX INTERPRETER

In Ref. [9], an infinite loop algorithm is proposed which runs within BDI agents. In this investigation, we adapt the existing algorithm and we create the "BDIx interpreter" process that runs on BDIx agents. The adapted algorithm shown in Algorithm 3.1 is based on the algorithm described in Ref. [9], however refined in such a way that it can be used in the BDIx agent environment. The adaptations involves a Plan Library, filtering of Sensor values, filtering and handling of Events, and execution of Intentions. Additionally, in order to execute a new iteration, the new BDIx interpreter waits for raised of events or change on sensor values targeting the reduction of the interpreter execution circles. Finally, it aims to handle the Desires and convert them to Intentions, as required in the DAI framework (see Figure 3.4). The specified algorithm is implemented in the agents' program.

Note that the Intentions in the BDIx interpreter algorithm at the line "intention-execute()" are executed concurrently and are limited to device CPU, memory, and battery power. To keep BDIx agents "light" in terms of execution, the BDIx Interpreter process through the Plan Library, limits the concurrent execution of Intentions to N (see Section 3.6.2). Additionally, the BDIx Interpreter can accept interruptions from events or changes on sensor values at any time of execution and can adjust the priorities values decision accordingly, by starting the iteration from the beginning. More specifically, the get-new-external-events and get-new-external-sensor-values are represented by two queues that hold events and values not currently taken under consideration.

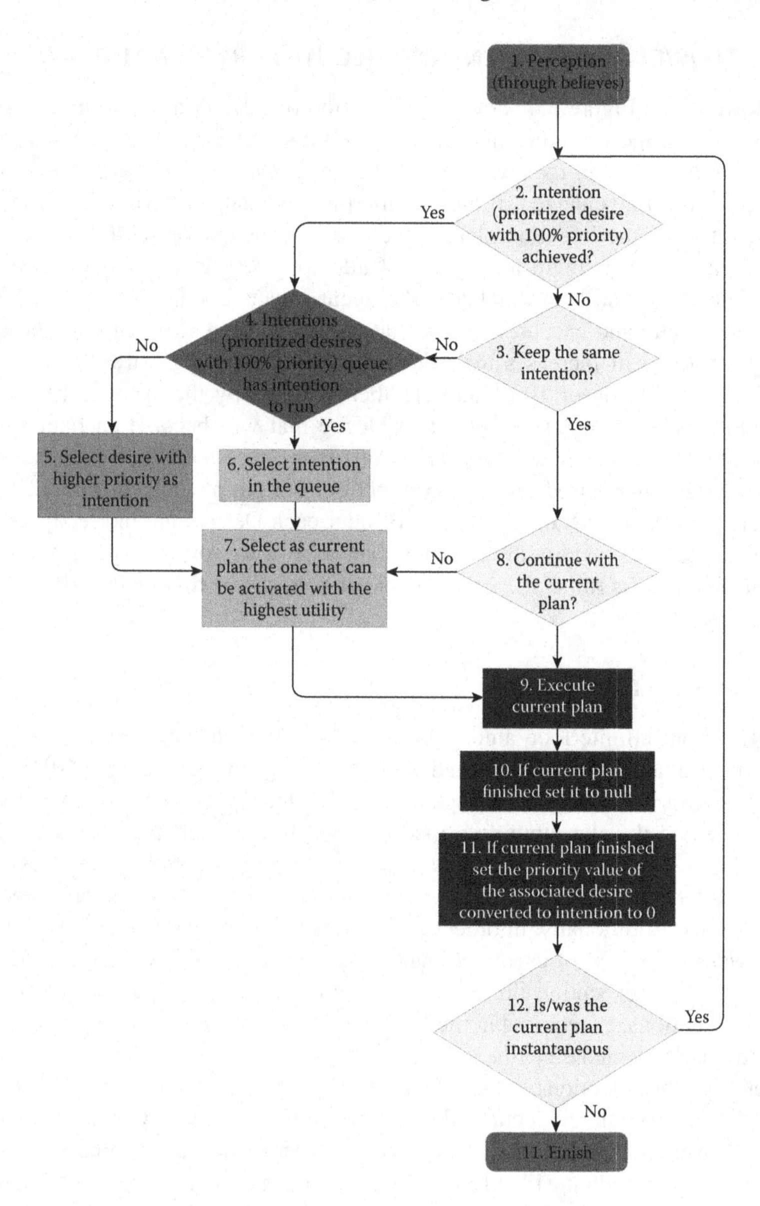

Figure 3.4 Flowchart of BDIx agent operation.

3.6.5 TELECOM OPERATORS AND THE DAI FRAMEWORK

With the use of BDIx agents and fuzzy logic planning, the DAI framework can be considered as a framework that is based on LEGO-based components. More specifically, the components of the framework are: (i) the Beliefs of the agent; (ii) the Desires of the agent; (iii) the plans that are associated with each Desire; (vi) the

Algorithm 3.1 BDIx˙Interpreter

```
1: PL:Planing Library
2: procedure BDIx˙INTERPRETER(()PL)
3:     do
4:         options: option-generator(event-queue,PL)          ▷ Planing Library needed parameters
5:         selected-options-desires: deliberate based on priority(options,desires,PL)   ▷ Deliberate with Plan Library
6:         intention:update-Intentions(selected-options,PL)
7:         goals:update-Goals(selected-options,PL)
8:         intention-execute(goals)   ▷ Run plans based on Intentions in the Goals queue with a call to the Intention
   Execution Algorithm shown in Section 3.6.3 for each Intention.
9:         event-queue:get-new-external-events,
10:        get-new-external-sensor-values and
11:        update Beliefs based on filters provided(PL)
12:        drop-unsuccessful-attitudes()
13:        drop-impossible-attitudes()
14:        remove-completed-intentions(update-priorities,make-them-desires,goals)   ▷ This is a fail safe check in a
   case, a Finished Intention is not removed from the Goals and it runs by the Intention Execution Algorithm.
15:        wait (until new external-events raised on event-queue or external-sensor-values changed)
16:        new˙events: new external-events raised on event-queue
17:        current˙external˙sensor˙values: current external-sensor-values
18:        old˙external˙sensor˙values: old external-sensor-values
19:    while (count(new˙events) > 0 || (current˙external˙sensor˙values ≠ old˙external˙sensor˙values))
20: end procedure
```

threshold values; (v) the events that the agent will react; and (vi) the Plan library that handles the priority of the Desires to become Intentions. All these six components can be changed at run-time by the telecom operators using API interfaces. In addition, these essential components can be added or removed at run time as long as the BDIx agent does not use them during the pursuing of Intentions. However, if they are used in Intentions, the agent can reset its states with an update.

3.7 DAI FRAMEWORK IN TERMS OF THE DAI IMPLEMENTATION, REQUIREMENTS AND CHARACTERISTICS

To start with, the DAI Framework can be implemented using predefined agent implementation frameworks (e.g., JADE, JASON) that support BDI agents and with the use of well-defined languages (e.g, Foundation for Intelligent Physical Agents - Agent Communication Language FIPA ACL) for messages exchange among them.

The DAI framework achieves the DAI requirements and characteristics (as shown in Section 2.1.1.2) with the utilization and the extension of BDI agents. Thus, in terms of requirements, the agent's granularity is acting at a task-level problem decomposition, the agents knowledge is specialized, the distribution of control is "shifting roles" because an entering UE can arrange the existing network and the message-model is used in high-level content. Additionally, in terms of characteristics, the proposed framework is a method of distribution of powers and communication of the agents. The agent architecture is homogeneous and reactive with filtering and reduced deliberation. Also, a message-model is used for the communication channel, the FIPA-ACL is used as a communication protocol and human is not involved in the decisions of the agent.

Thus, given the features of the DAI framework discussed in Section 3.2 and the characteristics shown in Section 3.6, I postulate that all properties of BDI agents remain valid for the BDIx agent. My assertion in based on the fact that the BDIx agents do not violate any rules of DAI and BDI agents (i.e. can execute perfectly parallel workload without any dependency or need for communication between those parallel tasks/nodes). Additionally, each BDIx agent utilizes only the LTE ProSE messages broadcast by the existing D2D-Relay devices that already execute the required actions to join the D2D communication network without depending on the decision from any other device. Finally, every BDIx agent executes predefined plans related to its Beliefs, Desires and Plan Library, and there is no need to have any coordination among them. Thus, the attributes of the system are maintained by the DAI design.

3.8 COMPUTATIONAL COMPLEXITY OF DAI FRAMEWORK

In terms of the computational complexity, the DAI framework uses the BDIx interpreter and the fuzzy logic controller in the Plan Library as shown in Sections 3.6.4 and 3.6.2. More precisely, the BDIx interpreter can perform n executions, where n is associated with the number of events raised or number of changes occurred on the sensors values. The aforesaid changes force the re-execution of the BDIx interpreter to consider any variations which occur in their surrounding environment. Every time the BDIx interpreter is executed, the integrated Plan Library uses fuzzy logic calculation to set priority values on the Desires, setting their order of execution based on the changes occurred. Calculating the computational complexity of fuzzy logic is beyond the scope of this book, however given the simplicity in the proposed fuzzy logic engine and its infrequent call we expect DAI computational load to be rather low (could be in the order of $O(n)$), especially if fuzzy logic is implemented as a table with parameters [23]. Moreover, in Ref. [24] there is an examination on fuzzy logic computational complexity, estimated to be $O(g(n))$. The complexity is separated in factors and is calculated to be $\theta(n.c.p)$, where n is the number of patterns, c is the number of clusters, and p is the dimension of the data points resulted to be $g(n)$. Based on the [24], the computational complexity of the DAI framework can be represented by $O(n.g(n))$.

REFERENCES

1. A. M. Schweidtmann and A. Mitsos, "Deterministic global optimization with artificial neural networks embedded," *Journal of Optimization Theory and Applications*, vol. 180, no. 3, pp. 925–948, 2019.
2. M. Bratman, *Intention, Plans, and Practical Reason*. Cambridge, MA: Harvard University Press, 1987.
3. M. Aschermann, P. Kraus, and J. Müller, "Lightjason: A BDI framework inspired by Jason," in *EUMAS/AT*, Valencia, 2016.
4. C. J. Amaral, "Embedding multi-agent system frameworks: A benchmarking," *Anais do Computer on the Beach*, vol. 9, pp. 939–941, 2018.
5. P. Novák and J. Dix, "Modular BDI architecture," *Proceedings of the International Conference on Autonomous Agents*, vol. 2006, pp. 1009–1016, Hakodate, May 2006.

6. B. Seok, J. C. S. Sicato, T. Erzhena, C. Xuan, Y. Pan, and J. H. Park, "Secure D2D communication for 5G IoT network based on lightweight cryptography," *Applied Sciences*, vol. 10, no. 1, p. 217, 2020.
7. A. Zhang, J. Chen, R. Q. Hu, and Y. Qian, "Seds: Secure data sharing strategy for D2D communication in LTE-advanced networks," *IEEE Transactions on Vehicular Technology*, vol. 65, no. 4, pp. 2659–2672, 2016.
8. V. Morreale, S. Bonura, G. Francaviglia, F. Centineo, M. Cossentino, and S. Gaglio, "Reasoning about goals in BDI agents: The PRACTIONIST framework," *CEUR Workshop Proceedings*, vol. 204, pp. 187–194, Utrecht, Feb. 2006.
9. A. S. Rao and M. P. Georgeff, "BDI agents: From theory to practice.," *ICMAS*, vol. 95, pp. 312–319, 1995.
10. A. G. Hernández, A. El Fallah-Seghrouchni, and H. Soldano, "Distributed learning in intentional BDI multi-agent systems," *Proceedings of the Fifth Mexican International Conference in Computer Science, ENC 2004*, pp. 225–232, Colima, Sep. 2004.
11. N. Babu Pamula, K. Chandra, and L. Kavitha, "Multi-agent system by using BDI in artificial intelligence," *International Journal for Development of Computer Science & Technology (IJDCST)*, vol. 5585, pp. 2321–338, 2016.
12. Y. Shoham and K. Leyton-Brown, *Multiagent Systems: Algorithmic, Game-Theoretic, and Logical Foundations*. Cambridge, MA: Cambridge University Press, 2008.
13. A. Guerra-Hernández, A. El Fallah-Seghrouchni, and H. Soldano, "Learning in BDI multi-agent systems," in *Computational Logic in Multi-Agent Systems* (J. Dix and J. Leite, eds.). Berlin, Heidelberg: Springer, pp. 218–233, 2005.
14. B. Subagdja and A. H. Tan, "Planning with iFALCON: Towards a neural-network-based BDI agent architecture," *Proceedings - 2008 IEEE/WIC/ACM International Conference on Intelligent Agent Technology, IAT 2008*, vol. 2, pp. 231–237, Sydney, Dec. 2008.
15. P. Caillou, B. Gaudou, A. Grignard, C. Q. Truong, and P. Taillandier, "A simple-to-use BDI architecture for agent-based modeling and simulation," in *Advances in Social Simulation 2015* (W. Jager, R. Verbrugge, A. Flache, G. de Roo, L. Hoogduin, and C. Hemelrijk, eds.). Cham: Springer International Publishing, pp. 15–28, 2017.
16. P. Caillou, B. Gaudou, A. Grignard, C. Q. Truong, and P. Taillandier, "A simple-to-use BDI architecture for agent-based modeling and simulation," *Advances in Intelligent Systems and Computing*, vol. 528, pp. 15–28, 2017.
17. V. M. Le, B. Gaudou, P. Taillandier, and D. A. Vo, "A new BDI architecture to formalize cognitive agent behaviors into simulations," in *7th Conference KES on Agent and Multi-Agent System: Technologies and Applications (KES-AMSTA 2013)*, Hue City, Vietnam, vol. 252, pp. 395–403, May 2013.
18. W. Jager, R. Verbrugge, A. Flache, G. de Roo, L. Hoogduin, and C. Hemelrijk, *Advances in Social Simulation 2015*, vol. 528. New York: Springer International Publishing, 2017.
19. M. Wiering and v. O. Martijn, *Reinforcement Learning: State-of-the-Art*. Berlin, Heidelberg: Springer, 2013.
20. J. L. . Feliu, "Use of Reinforcement Learning (RL) for plan generation in Belief-Desire-Intention (BDI) agent systems," Master's thesis, University of Rhode Island, 2013.
21. P. Lokuge and D. Alahakoon, "Reinforcement learning in neuro BDI agents for achieving agent's intentions in vessel berthing applications," *Proceedings of the International Conference on Advanced Information Networking and Applications, AINA*, vol. 1, pp. 681–686, Taipei, Mar. 2005.
22. T. Haerder and A. Reuter, "Principles of transaction-oriented database recovery," *ACM Computing Surveys (CSUR)*, vol. 15, no. 4, pp. 287–317, 1983.

23. Y. H. Kim, S. C. Ahn, and W. H. Kwon, "Computational complexity of general fuzzy logic control and its simplification for a loop controller," *Fuzzy Sets and Systems*, vol. 111, no. 2, pp. 215–224, 2000.
24. E. P. Ephzibah, "Time complexity analysis of genetic- fuzzy system for disease diagnosis," *Advanced Computing: An International Journal*, vol. 2, pp. 23–31, 2011.

Notes

[1] An introduction to the DAI framework and BDI agents appears in Chapter 2.
[2] In DAI framework classification (see Section 3.3), the term used is decentralized. We adopted distributed as it better conveys the implemented nature of the proposed DAI framework.
[3] These languages achieve agent intercommunication and are designed for BDI agents with the target of solving problems collaborative or exchanging information.
[4] For example, the agents can use the IP address of BDI agents, shared among UEs over LTE Proximity Services.
[5] Autonomicity: Having the freedom to act independently in order to solve a problem.
[6] Dynamicity: Characteristic of the approach to react to changing conditions of operation (e.g., a D2D device changes coordinates, increases speed, etc.) and continue satisfying the D2D Challenges. This is also called Dynamic Implementation at features.
[7] Flexibility: Ability to adapt to possible, future changes in its requirements (e.g., increase the number of devices, add mmWaves, D2DSHR go offline) and react fast in a change of a situation (e.g., a D2D device enters/leaves the D2D network)
[8] For clarity, we will use D2D-Relay to represent both the direct hop (D2DSHR) and the multihop relay (D2DMHR) cases.
[9] Empirically $M = 10$ is a good choice because it enables the plan to finish by M retries. Again the number is set due to authors experience in parallel programming at android development.
[10] The number 10 for concurrent executions of Intentions-Plans in the BDIx agent is set empirically from authors experience at parallel programming on android.
[11] ACID is a set of properties of database transactions intended to guarantee data validity despite errors, power failures, and other mishaps. In the context of databases, a sequence of database operations that satisfies the ACID properties (which can be perceived as a single logical operation on the data) is called a transaction (e.g., write the correct Data Rate after changing position could result as an atomic transaction).

4 Implementation Specifics of the DAI Framework for D2D Communication

To better illustrate the concepts of the DAI framework described in Chapter 3, in this chapter, we will consider a device-to-device setup. In this setup, each D2D device aims to tackle the D2D challenges by focusing on the local environment of D2D communication. Additionally, implementation specifics about the DAI framework and BDIx agents for D2D communication are discussed and provided in this chapter. This chapter shows how the DAI framework can address the D2D challenges with the use of BDIx agents featuring customized Beliefs, Desires, a Plan Library, and Fuzzy Logic. Additionally, it promotes the idea that the D2D communication is not a global problem. Therefore, it should not be handled as a global problem that must be solved by the Base Station, but as a problem that should be addressed in a distributed fashion with the use of artificial intelligence. Hence, this book proposes that the control is handled locally by the UEs, in order to form communication links in a shorter time [1–9] and establish more effective D2D communication. This book considers that the use of Distributed Artificial Intelligence (DAI) [10–12] control is most suitable for this challenging and dynamic environment of D2D communication.

This chapter provides the following: Firstly, the BDIx agent setting in the DAI framework used for D2D communication is described. Then, potential metrics that can be used by the BDIx agent in its plans along with the implementation constraints that the BDIx agent must also consider are discussed. After that, focus is given on the implementation specifics aspects for D2D communication. Moreover, technologies that can used by the BDIx agents are also investigated. Continuing, the implementation of the DAI framework Fuzzy Logic controller in D2D communication is provided. Moreover, the implementation of reinforcement learning in BDIx agent is shown at the D2D communication. Finally, a comparison is executed among the existing D2D frameworks with the DAI framework.

4.1 BDIX AGENT SETTINGS IN DAI FRAMEWORK FOR D2D COMMUNICATION

In this section, the BDIx agent components and how their behaviour is realized in the DAI framework are described. Also, their importance for the DAI framework is discussed.

The BDIx agent is characterized by the components it comprises (Belief, Desire, Intentions and Goals) and its behaviour (Perception, Plan):

DOI: 10.1201/9781003469209-4

1. The BDIx agent components are realized in the DAI framework, as follows: (i) The Beliefs can be prolog like facts, variables, or any other data structure (i.e., neural networks); (ii) The Desires, Intentions, and Goals can be a list, a stack, or a queue;
2. The BDIx agent behaviour is realized in the DAI framework, as follows: (i) The perception is a part of the planning library and it is realized with the use of fuzzy logic neural networks [13] and IF-THEN statements; and (ii) The plans are realized with the use of any programming language (e.g., Java, C++, Python) as methods.

Note that Beliefs, Desires, and Intentions of a BDIx agent can be changed/extended, at any time and on the fly, according to changes affecting the D2D network structure or based on raised events. The events that will be raised can be pre-specified, either with the declaration of thresholds (i.e., constant variables that when exceeded an event can be raised) or with specific network events. Through the raised events, the BDIx agent can achieve re-enforcement learning and react in a specific manner towards achieving its tasks. This is a flexibility offered by the proposed DAI framework. Concentrated on the aspect of D2D communication, an indicative list of events that can be raised as well as a list of Beliefs and Desires that the agent must have with the purpose of joining/creating a D2D communications network are provided in Table 4.1 for Events, Table 4.2 for Beliefs and Table 4.3 for Desires.

Table 4.1
Events

Events
Power monitor issue (battery power level reduced below)
Security monitor issue (security breach)
Dense network (due to bandwidth utilization number of UEs under D2DSHR)
UE enters/leaves the D2D network
D2D move away from AP (D2D-relay)
Shift D2D UE(s) to other D2D-relay
QoS & QoE issue due to data rate or signal quality reduction
QoS & QoE issue due to distance to AP
QoS & QoE issue due to dense network

4.2 POTENTIAL METRICS TO BE USED BY THE BDIX AGENT

The basic measurement values (metrics) that can be used in the plans executed by the BDIx agent addressing the Desires associated with the D2D challenges, are: (i)

Table 4.2
Beliefs

Beliefs
Frequency band connected to BS
Battery power level
Used metric value (e.g. Weighted Data Rate (WDR) as shown in Section 6.1.2, Channel Quality Indicator (CQI), interference)
Transmission mode selected
Frequency band used
Best reused frequency band to be used with less interference.
% of bandwidth utilization
Data rate
Lat/long (coordinates)
Number of D2D devices in D2D network
Next hop that D2D device connects to (from D2D-relay at D2DMHR/BS and D2DC at D2DSHR)
Distance from the next hop that the D2D device (UE) connect to
coordinates of the next hop that the D2D device connects to
% signal quality to where I connect to
% data rate change to where I connect to
speed (D2D device moving speed)
Number of users that the D2D device serves (if transmission mode is D2DSHR)
IPs/MSISDN of users that the D2D device serves (if transmission mode is D2DSHR)
Sharing subnet (if transmission mode is D2DSHR)
IP v4
IP v6
Deep neural network (DNN[1]) to calculate underlay frequencies to be used
List of D2DSHRs with coordinates, frequency band, number of D2D clients serve, frequencies shared to D2D clients (inband or outband) and metric used (e.g. WDR)
List of D2DMHRs with coordinates, frequency band, frequencies shared to D2D clients (inband or outband) and metric used (e.g. WDR)
Round time of packet to access gateway
Number of D2D clients that the D2D device serves as D2DSHR
Security breach
Counters of packets for each D2D client (for security reason)
Fuzzy logic (IF-THEN rules) assigning priority values on the desires based on events and beliefs
Transmission power
Back-propagation neural network used in Section 3.4

Table 4.3
Desires

Desires
Preferred network is D2D network always with 100% priority
Hardware health is acceptable
Identify the surrounding D2DSHRs and D2DMHRs
Find the best reused frequency with the least interference
Find best transmission mode that achieves the best achievable
Signal quality, data rate and WDR
Signal quality is acceptable
Data rate is acceptable
Used metric value (e.g. WDR shown in Section 6.1.2) is acceptable
Achieve maximum sum rate
Distance of D2D client device with D2DSHR is acceptable
Number of D2D client that the D2D device serves as D2DSHR is acceptable
Bandwidth consumed by Users that the D2D device serves as D2DSHR is acceptable
Achieve QoS specified by 5G requirements, always with 100% priority
Achieve QoE specified by 5G requirements, always with 100% priority
The latency (round time/ultra-reliable low latency communication) of accessing gateway or any other D2D device is acceptable, always with 100% priority
Battery power level reservation at D2D device, always with 100% priority
Security monitoring at D2D device, always with 100% priority

link data rate; (ii) total sum rate; (iii) link SINR; (iv) link signal-to-noise-ratio (SNR); (v) link power; (vi) QoE/QoS of UE; (vii) link spectral efficiency; and (viii) weighted sum rate (as shown in Ref. [14]); (ix) path WDR (as described in Section 6.1.2 and explained below); (x) total power; (xi) total spectral efficiency; (xii) link interference; and (xiii) location coordinates.

In this illustrative example, the weighted data rate (WDR) is adopted and used by the BDIx agent to execute optimized control in its cellular (i.e., LTE, 5G) and Wi-Fi interfaces. It is defined as the data rate of the weakest link in a path that the device is connected. The WDR as a metric represents the network paths towards a BS by a value. When a D2D device enters the network, this value is acquired from each neighbouring D2D, D2DSHR, and D2DMHR Device for each path they can potentially belong to by message exchanges with their neighbours with the use of LTE ProSe. With WDR, the agent is using only local environment information (e.g., the coordinates of the D2D devices in proximity), rather than the global environment, with the use of LTE ProSe. With the LTE ProSe, each D2DSHR and D2DMHR can share its WDR with the other D2D devices. Thus, by relying only on the local environment, reduced signalling overhead and much faster control decision-making are expected.

4.3 IMPLEMENTATION CONSTRAINTS OF DAI FRAMEWORK

Due to the nature of the wireless networking problems that the implemented DAI framework tries to tackle, there are some realization constraints that must be considered in the design of the DAI framework. These are the following:

- The location (lat/long) of the D2D device must always be known (e.g. through GPS) by the BDIx agent that resides in it.
- Each D2D device should have a BDIx agent installed in a secure memory place, that only the network operator can access (e.g. over REST APIs) and manipulate.
- The BDIx agent can query the information of other neighbouring cells only through the cell that the D2D device resides. In a case of a disaster recovery, the information needed to be queried and related to the neighbouring cell is: (i) signal strength; (ii) coordinates; and (iii) identification number (cellid).
- The D2D device in the DAI framework should at least have one mobile interface and one Wi-Fi interface. The mobile interface is needed for the establishment of links to BS, D2DSHR and D2DMHR D2D devices. The Wi-Fi interface is needed for establishing links to the D2DSHR CH (when the Device is D2D Client) or to the Wi-Fi Gateway (when Device is D2DMHR).
- In the DAI framework, all UE devices under the cell are free to enter the D2D communication network and be able to operate as D2D devices or stay connected as it is to the BS. However, for the D2D devices that decided to stay connected to the BS, they should select the D2DMHR transmission mode.

4.4 IMPLEMENTATION SPECIFICS FOR MEETING THE D2D REQUIREMENTS/CHALLENGES WITHIN THE DAI FRAMEWORK

In this section, we investigate how the proposed DAI framework can successfully meet the D2D challenges (see Section 4.4.1). More specifically, the relation among the network events, thresholds and D2D challenges with the Beliefs and Desires is illustrated. The impact of the network events and the Plan Library on the calculation of the Desires' priority values is also provided. Additionally, the association and dependency rules among D2D challenges/requirements are described. Finally, to reduce unnecessary Intention executions by the BDIx agent, a mechanism for filtering the most valuable events is implemented.

4.4.1 REALIZATION OF D2D CHALLENGES WITHIN A BDIX AGENT ENVIRONMENT

In order to actualize D2D communication and accomplish 5G requirements, several D2D challenges as shown in Section 2.2.2 need to be addressed. These include device discovery (DD), frequency mode selection (FMS), transmission mode selection (TMS), interference management (IM), power control (P-C), security (S), radio resource allocation (RRA), Cell Densification & Offloading (CDO), QoS & QoE (QoS/QoE), use of mm-wave communication, non-cooperative users (NCU), and handover management (HO).

Within a BDIx agent environment, D2D challenges are defined as requirements and indirectly as Desires with the purpose to be realized by the BDIx agents. The D2D challenges are implemented with the use of the plans that are associated with the related Desires. In addition, some D2D challenges must be handled when specific network events are raised (i.e., when a device is entering the D2D network), and some Beliefs can change due to sensors readings or events raising. Tables 4.4 and 4.5 describe how all the aforesaid are associated.

The D2D challenges related to the utilization of mm-waves in D2D communication, the NCU, and device discovery are special cases that can be handled as shown below:

The inclusion of mm-waves for the establishment of a D2D link is decided during the transmission mode selection process, using prior information acquired through ProSe messages during the device discovery process.

Non-cooperative users (NCU) are not considered in our DAI framework, as this is not allowed by the present form of the BDIx agents. All devices joining the network are forced by the installed BDIx agent to cooperate in the achievement of the D2D challenges. Future extensions could be investigated to also consider NCUs.

The Device Discovery can be solved with the use of LTE ProSe existing technology.

4.4.2 D2D CHALLENGES: INTER-DEPENDENCY, ASSOCIATION AND PARTIAL IMPLEMENTATION RULES

An important concept related to the implementation of the D2D challenges, which are translated into BDIx agent Desires, is their interdependency (i.e., there are D2D challenges whose achievement depends on the successful achievement of some others). For example, in order for "Transmission Mode Selection" to be achieved, device discovery must be completed first. If there is a dependency between two D2D challenges, a dependency rule is defined and reflected in the Plan Library by setting the fuzzy logic rules (IF-THEN statements) defining their priority of execution. This will guarantee that the Desires (i.e., D2D challenges) that are currently under the active pursuit of the agent (i.e., running as Intentions in Goals) do not have any other interdependency (i.e., keep ACID property; see Section 3.6.2) and thus will be successfully finalized.

Table 4.4
Relation among Network Events, BDIx Agents Events (Raised by Thresholds), D2D Challenges and Desires

Network Events	BDIx Agent Defined Events (Raised from Thresholds) Associated with Network Events	D2D Challenge Associated with the Events & Order of Execution of D2D Challenges	D2D Desires Associated with D2D Challenges
D2D device enters a D2D communication network	UE is entering the mobile network	DD, IM, RRA FMS, TMS	Identify the surrounding D2DSHRs and D2DMHRs (DD), Find the best-reused Frequency Band with the least interference (IM + RRA), Signal quality is acceptable (FMS + TMS), Data Rate is acceptable (FMS + TMS), Speed that D2D device is moving is acceptable (D2DSHR or D2DMHR) 1.5 m/s (pedestrian) (FMS + TMS), Battery power of D2D device is acceptable (D2DSHR or D2DMHR) (FMS + TMS), Distance of D2D device with D2D-Relay is acceptable (FMS + TMS)
D2D device moves away from D2D-Relay	Minimum signal quality accepted. Minimum data rate accepted. The maximum speed that node is moving in order to be D2D. D2DSHR or D2DMHR 1.5 m/s (pedestrian). Minimum signal quality drop. Maximum distance to move away from a D2D-Relay.	DD, IM, RRA FMS, TMS, HO	The same as above
Always runs, does not require an event to run	Runs at all times. The desire is always the intention with 100% priority.	S	Security monitoring at D2D device (S)

(Continued)

Table 4.4 (*Continued*)
Relation among Network Events, BDIx Agents Events (Raised by Thresholds), D2D Challenges and Desires

Network Events	BDIx Agent Defined Events (Raised from Thresholds) Associated with Network Events	D2D Challenge Associated with the Events & Order of Execution of D2D Challenges	D2D Desires Associated with D2D Challenges
D2D device wants to join an already dense network	Maximum number of users supported by CH (D2DSHR).	CDO, HO	Number of users that the D2D device serves as D2DSHR is acceptable (CDO + HO), Bandwidth consumed by users that the D2D device serves as D2DSHR is acceptable (CDO), The same Desires that are associated with the Mode Selection section above (HO).
QoS, QoE not achieved, runs always (with BPNN shown in Section 3.4)	Minimum signal quality accepted. Minimum data rate accepted. Minimum signal quality drop. Maximum distance to move away from a D2D-Relay.	QOS/QOE, CDO, HO	Achieve QoS specified by 5G requirements (QOS), Achieve QoE specified by the user according to current and historical records (QOE), The latency (round time/ultra-reliable Low latency communication) of accessing gateway or any other D2D device is acceptable (QOE+ QOS),
Battery power reduced a lot in the D2D device more than a pre-specified threshold	Minimum battery level threshold exceeded	P-C	Battery power reservation at D2D device.

Table 4.5
Relation among Network Events, Thresholds, D2D Challenges and Beliefs

Network Events	BDIx Agent Defined Events (Raised from Thresholds) Associated with Network Events	D2D Challenge Associated with the Events & Order of Execution of D2D Challenges	D2D Desires Associated with D2D Challenges
D2D device enters D2D communication network	UE is entering the network	Frequency band connected to BS,Speed, List of D2DSHRs and Multi Hop Relays (D), Number of D2D devices in D2D network (D), Frequency band used (IM + RRA), Best reused frequency band to be used with less interference (IM + RRA), Transmission Mode Selected (FMS + TMS), WDR (FMS + TMS), Next Hop that D2D device connects to From D2D-relay at D2DMHR/BS and D2DC at D2DSHR) (FMS + TMS), Distance from the next hop that the D2D device (UE) connect to (FMS + TMS), Coordinates of the next hop that the D2D device connects to (FMS + TMS), n% signal quality to where I connect to (FMS + TMS), n% Data rate change to where I connect to (FMS + TMS), Signal quality to where I connect to it (FMS + TMS), % Data rate change to where I connect to it (FMS + TMS), Data rate (FMS + TMS), % of Bandwidth utilization (FMS + TMS)	
D2D device moves away from D2D-relay	Minimum signal quality accepted, Minimum data rate accepted, Maximum speed that node is moving in order to be D2DSHR or D2DMHR 1.5 m/s (pedestrian), Minimum signal quality drop, Maximum distance to move away a D2D-relay	same as above	

(Continued)

Table 4.5 (*Continued*)
Relation among Network Events, Thresholds, D2D Challenges and Beliefs

Network Events	BDIx Agent Defined Events (Raised from Thresholds) Associated with Network Events	D2D Challenge Associated with the Events & Order of Execution of D2D Challenges	D2D Desires Associated with D2D Challenges
QoS, QoE not achieved, Always runs, does not require an event to run	Minimum signal quality accepted, Minimum data rate accepted, Minimum signal quality drop, Maximum distance to move away from a D2D-relay	Same as above	
QoS, QoE not achieved, Dense network	QoS or QoE not achieved with QoE & QoS BPNN shown in Section 3.4	Frequency band connected to BS,Speed, List of D2DSHRs and D2DMHRs (D), Number of D2D devices in D2D network (D), Next Hop that D2D device connects to (From D2D-relay at D2DMHR/BS and D2DC at D2DSHR) (HO), Distance from the next hop that the D2D device (UE) connect to (HO), Coordinates of the next hop that the D2D device connects to (HO), n% signal quality to where I connect to (HO), n% Data rate change to where I connect to (HO), Signal quality to where I connect to it (HO), % Data rate change to where I connect to it (HO), Data rate (HO), % of bandwidth utilization (HO)	
D2D device wants to join an already dense network	Maximum number of users supported by CH (D2DSHR).	Same as above	
Battery power reduced a lot in the D2D device more than a pre-specified threshold	Minimum battery level threshold exceeded	Transmission power (P)	
Always runs, does not require an event to run - Security	All time (does not need an event). The Desire is always Intention with 100% priority	Security breach (S)	

A second important concept related to the D2D challenges is their association with a common Desire. For example, "Transmission Mode Selection" and "Frequency Mode Selection" are associated with the same Desire and are thus successfully achieved using the same plan. If there is an association between two D2D challenges, this is reflected in the Plan Library through the definition of association rules by a common Desire.

A third important concept related to the D2D challenges is their partial implementation by other D2D challenges (see Figure 4.1). For example, "Handover" of D2D device cannot be fully implemented by executing only the plan associated with it. For complete implementation, it is also partially implemented by "Mode Selection" and partially by "Cell Densification & Offloading". If there is a partial implementation between D2D challenges, this is reflected in the Plan Library through the definition of partial implementation rule(s) in conjunction with the raised event.

In Figure 4.1, we provide a graph showing the interdependency, association and partial implementation between D2D challenges: (i) Interdependency is depicted using ping lines with a direction depicting the dependency. As shown, we can have multiple levels of dependencies according to how many D2D challenges have to be completed beforehand in a specific order and path direction; (ii) Association is represented with two-way red arrows, and the associated D2D challenges are in a light orange box; and (iii) partial implementation is illustrated with one way light blue arrow line. Additionally, as shown in Figure 4.1, there are multiple arrow paths that conclude at "Handover" or "QoS&QoE" D2D challenges that realize the "partial implementation" rule. With the aforementioned arrow paths, there are dependency arrows and association boxes that show the rules of the D2D challenges that partially implement them. Also, the paths show the specific cases and the execution order on specific events.

An example that covers all the rules described above and depicted in Figure 4.1 is given below. For example, in the case of the "QoS, QoE issue due to Data Rate, Signal Quality Reduction" event, shown as the "1.1.1 Dependency flow" dependency line, the order of execution of each D2D challenge is: DD 100%, RRA & IM 99% with association rule, FMS & TMS 98% with association rule, with this execution the D2D challenge "QoS & QoE" is partially achieved, then the "QoS&QoE" gets 97%. Please note, for a different event, the order for the D2D challenge "QoS & QoE" could be different along with the execution path.

4.4.3 FILTERING OF EVENTS RAISED DUE TO THE PERSISTENT BEHAVIOUR OF BDIX AGENT

In order for the framework to tackle the problematic and computationally expensive behaviour (as described in Section 3.6.1), a filtering algorithm is implemented and used by the BDIx agent. The aim of the filtering algorithm is to filter out the unnecessary feedback or unnecessary raised events (see Figure 3.3) and avoid any unnecessary executions of plans by the BDIx agent. For example, in our DAI framework, an event is raised only when:

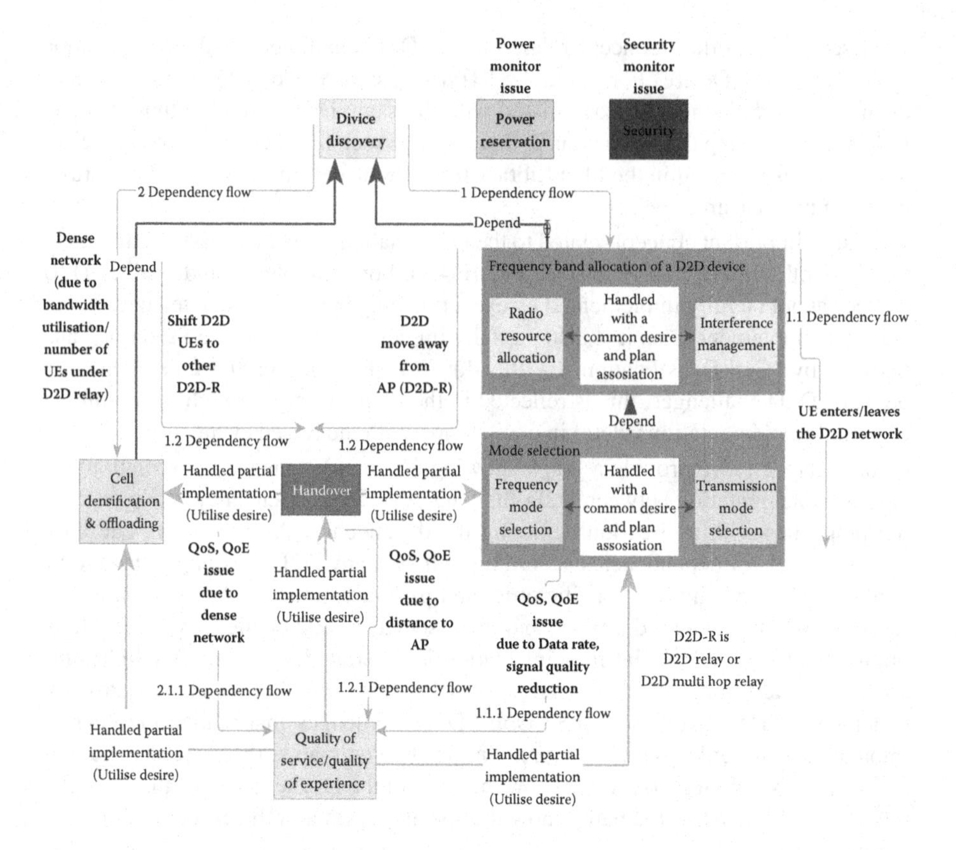

Figure 4.1 The D2D challenges relations.

The D2D device moves outside from the D2DSHR (CH) coverage.
The D2D device moves outside from the D2DMHR range.
The signal quality drops below 10%.
The battery level of the D2D device drops below 50%.
The number of D2D clients served by a D2DSHR becomes more than 200. In this case, the D2DSHR is considered as overloaded.

4.5 TECHNOLOGIES THAT CAN BE USED BY THE BDIX AGENTS

This section shows the technologies that can be used by the BDIx agent. Additionally, it provides for each technology any constraints (ended up to be threshold values) that the BDIx agent must consider before utilizing it.

The technologies that can be used by the BDIx agents, mainly for device discovery and intercommunication, are provided below:

- LTE ProSe technology can be utilized in order to execute device discovery. The proximity service (ProSe[2]) introduced in the 3GPP standards Release 12 until 14 is located in the Evolved Packet Core and allows devices to discover other peer devices in their proximity for D2D communication services.

- The Common Pilot Channel (CPICH) or the Physical Uplink Shared Channel (PUSCH) can be used by the agents for executing device discovery, communication, and exchange of messages.
- LTE Direct (Standard; not fully tested). This is a long-range and multi-user technology. In order to directly communicate and exchange data with LTE Direct, the neighbouring devices follow two phases for link establishment: (i) in the first phase, with the use of LTE ProSe, the "to be connected" devices need to send a registration message to the eNB with a ProSe application ID. Then the eNB organizes the communication between the devices using the control channels [15]; (ii) in the second phase the "to be connected" devices agree on a channel to be used along with the radio resource parameters and start to communicate.
- Wi-Fi Direct (Standard; well-tested and supported in latest mobile devices). Wi-Fi Direct is built upon the Wi-Fi technology. A Wi-Fi Direct device can share its link with other Wi-Fi-enabled devices. These devices can connect directly to it with an easy setup and discovery feature (with the use of service discovery) [16]. The standard uses the following secure protocols for communication: (i) Wi-Fi protected setup, and (ii) Wi-Fi protected access. Note, however, that a limited number of clients can connect to the sharing link (a maximum of 200 UEs can be under each cluster head).
- IP communication technology can be used by the BDIx agent to communicate with other agents. When a D2D candidate requests to enter the D2D communication network, it is already connected to the BS with inband Frequency Band (given by the BS) and already has an IP address. Hence, it can communicate with the other BDIx agents via IP. For a BDIx agent to communicate with other agents via IP, it needs to use a Device Discovery technology (as shown above, e.g., LTE ProSe) to acquire the information and IPs of the surrounding BDIx agents and communicate with them using agent language. Regarding IP, worth noting: (i) when a D2D device selects the D2D client transmission Mode, it will use a private IP subnet assigned by the D2DSHR; and (ii) when a D2D device selects the D2DSHR or D2DMHR transmission mode and it shares its bandwidth with a selected protocol (Wi-Fi Direct or LTE Direct) among its D2D devices, it will use the IP provided by the BS.

Note, for device discovery, in order to achieve DAI and the execution of the parallel tasks in UE device, we utilize the LTE ProSe messages/ share channels for the purpose not to have dependencies on other devices execution and decision. In this way, we do not have any results dependencies, and we isolate each execution of the parallel tasks with parallel control with the use of shared information. Thus, with DAI, we exploit extensive scale computation and spatial distribution of computing resources, and each node does the control in parallel. The intelligent agent approaches can only support this type of control.

In the DAI framework, both LTE direct and Wi-Fi direct technologies can be used by the BDIx agents to: (i) form D2DMHR connections between each D2DSHR towards BS and (ii) form D2D clusters (with a cluster head to be a D2D relay hop). The communication among BDIx agents who selected a transmission mode (i.e., D2D Client, D2DSHR, and D2DMHR) depends on how they access the network towards the Gateway and how they share the resources (i.e., over Wi-Fi with outband D2D or mobile network with inband D2D).

4.5.1 TECHNOLOGY SPECIFIC CONSTRAINS

Based on the technology used by the BDIx agent (i.e., Wi-Fi Direct, LTE Direct, and LTE ProSe; see Section 4.5), for achieving D2D communication, there are some specific technology constrains (as shown in Refs. [16–19]), that should be considered. More specifically, for the D2D-Relay:

- When Wi-Fi Direct is used: (i) Maximum number of D2D devices that can be supported is 200; and (ii) Maximum distance allowed among the D2D devices is 200 m.
- When LTE Direct[3] is used: (i) Maximum number of D2D devices that can be supported is 2000 and (ii) maximum distance allowed among the D2D devices is 600 m;
- When LTE proximity services are used the usage of PUSCH for device discovery is restricted.

To implement the technology constraints, constant values are used in the present implementation. These are saved in the BDIx agent threshold & constrains list (in a dedicated memory location).

4.6 IMPLEMENTATION OF THE DAI FRAMEWORK FUZZY LOGIC CONTROLLER IN D2D COMMUNICATION

In this section, we provide the implementation of the fuzzy logic controller for DAI framework which selects cases (Section 4.4.2, Figure 4.1) and indirectly assigns priority values for specific Desires. The inputs (antecedents), outputs (consequent), knowledge base, and the linguistic inference system (Fuzzy Inference System/Engine) which are applied in the process of fuzzyfication, fuzzy inference system control, and defuzzyfication are discussed below and pictorially viewed in Figure 4.2.

Note that the implementation of the fuzzy logic is executed when the threshold values are exceeded or specific events are raised (Section 4.4.1, Tables 4.4 and 4.5).

4.6.1 ANTECEDENTS

The values of the following input variables with their respective range at universe, membership functions, and linguistic variables (set empirically) are used for fuzzification:

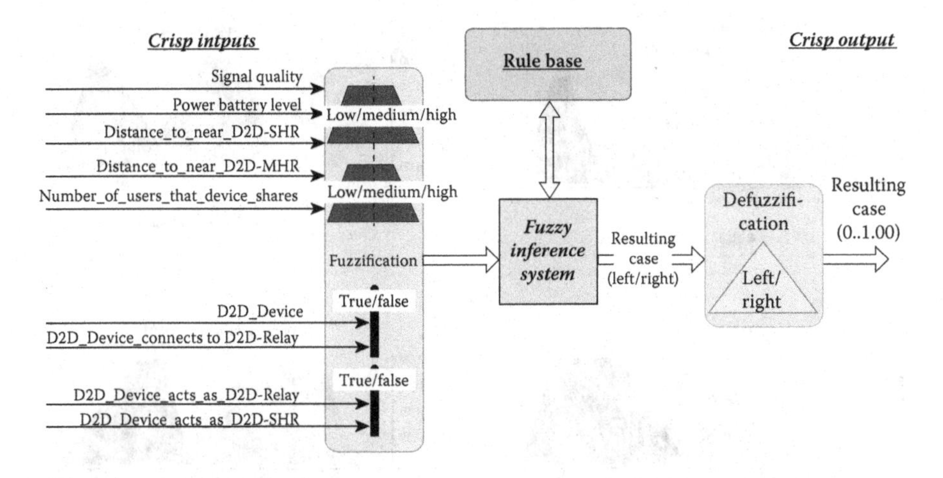

Figure 4.2 Fuzzy logic implementation for DAI framework.

Signal Quality (0...100)%. The selected membership function is trapezoidal and takes three values at Fuzzy Set: low (0, 15, 25, 45), medium (30, 45, 55, 85) and high (75, 85, 90, 100). The representation of the membership function is shown in Figure 4.3a.
Power Battery Level (0...100)%. The selected membership function is trapezoidal and takes three values at Fuzzy Set: low (0, 15, 25, 35), medium (30, 45, 55, 75) and high (65, 85, 90, 100). The representation of the membership function is shown in Figure 4.3b.
Distance˙to˙near˙D2DSHR (0...200). The selected membership function is trapezoidal and takes three values at Fuzzy Set: low (0, 15, 25, 60), medium (50, 70, 100, 120) and high (100, 140, 180, 200). The representation of the membership function is shown in Figure 4.3c.
Distance˙to˙near˙D2DMHR (0...500). The selected membership function is trapezoidal and takes three values at Fuzzy Set: low (0, 15, 25, 60), medium (50, 100, 190, 360) and high (300, 340, 380, 500). The representation of the membership function is shown in Figure 4.3d.
Number˙of˙users˙that˙device˙shares (0..200). The selected membership function is trapezoidal and takes three values at Fuzzy Set: low (0, 55, 75, 110), medium (90, 120, 130, 180) and high (160, 170, 180, 200). The representation of the membership function is shown in Figure 4.3e.
Fuzzy singletons, that for input "1" return the value of "true" and for input "0" return the value of "false":

- D2D˙Device (if the device is part of the D2D network).
- D2D˙Device˙connects˙to˙D2D-Relay (if the device is part of the D2D network).
- D2D˙Device˙acts˙as˙D2D-Relay (if the device is part of the D2D network).
- D2D˙Device˙acts˙as˙D2DSHR (if the device is part of the D2D network).

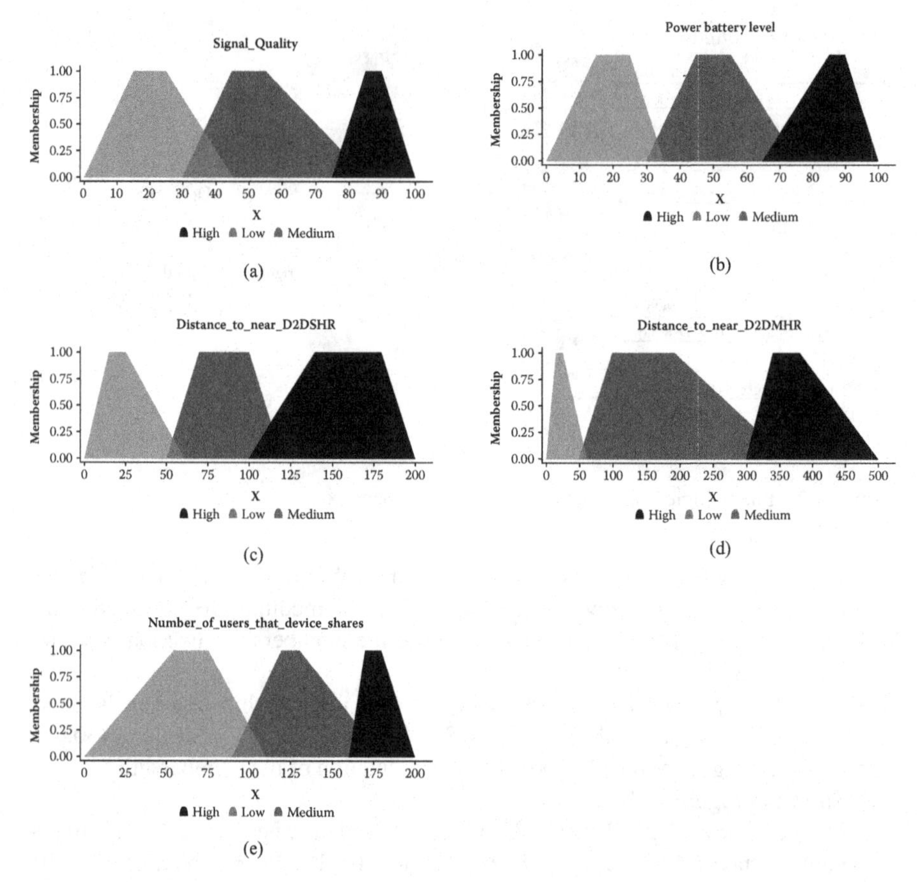

Figure 4.3 Membership functions. (a) Membership function for signal quality. (b) Membership function for power battery level. (c) Membership function for distance˙to˙near˙D2DSHR. (d) Membership function for distance˙to˙near˙D2DMHR. (e) Membership function for distance˙to˙near˙D2DMHR.

4.6.2 CONSEQUENT

As shown in the interdependency, association and partial implementation between D2D challenges in Figure 4.1, there are two cases of execution of Intentions. In the first case, the fuzzy logic should decide the left branch of the tree if there is an event or threshold value related to the "Cell Densification and Offloading" for Handover. In the second case, the fuzzy logic should decide the right branch of the tree if there is an event or threshold value related to the "Mode Selection" for Handover. Then after deciding the case, the Plan Library will set the priority values of the Desires.

The values of the "Resulting Case" output variable with its respective range at universe, membership function and linguistic variables (set empirically) used for defuzzification are:

The defuzzification, membership function is triangular and takes two values at fuzzy set, left (0,0.25,0.50) and right (0.50,0.75,1.00). The final output value (resulting case) after defuzzification are crisp with decimals from 0 to 1.00. The Plan Library will choose the correct case according to the resulting crisp output value as follows:

- <0.50 selects the Left case in the branch of the tree. The "left" case assigns priority values of 99% to "Cell Densification and Offloading" and 100% to "Handover".
- ≥0.50 selects the Right case in the branch of the tree. The "right" assigns priority values of 98% to "Mode Selection", 99% to "Radio Resource Allocation and Interference management" and 100% to "Device Discovery".

4.6.3 RULES USED IN THE LINGUISTIC INFERENCE SYSTEM

In this examination, we selected the "Mamdani Fuzzy Model" due to the simplicity of its design. The following rules are selected:

- If D2D˙Device is "false", then the Resulting Case is "right". *In this case, the device enters the network.*
- If D2D˙Device is "true" and D2D˙Device˙acts˙as˙D2D-Relay is "false", and Signal Quality is "low" then the Resulting Case is "right". *In this case, the device has bad signal quality (moves away from the D2D-Relay).*
- If D2D˙Device is "true" and D2D˙Device˙acts˙as˙D2D-Relay is "false", and Distance˙to˙near˙D2DSHR is "far", then the Resulting Case is "right". *In this case, the device moves away from the D2DSHR as D2D Client.*
- If D2D˙Device is "true" and D2D˙Device˙acts˙as˙D2D-Relay is "true", and Distance˙to˙near˙D2DMHR is "far", then the Resulting Case is "right". *In this case, the device moves away from the D2DMHR as D2D Single Hop Relay or as D2D Multi-Hop Relay.*
- If D2D˙Device is "true" and D2D˙Device˙acts˙as˙D2D-Relay is "true", and Signal Quality is "low", then the Resulting Case is "right". *In this case, the device has bad signal quality (moves away from the D2D-Relay) as D2D Single Hop Relay or as D2D Multi-Hop Relay.*
- If D2D˙Device is "true" and D2D˙Device˙acts˙as˙D2D-Relay is "true", and Power Battery Level is "low", then the Resulting Case is "left". *In this case, the device has low battery, and it needs to shift all D2D Devices that connect to it t other D2D-Relay.*
- If D2D˙Device˙acts˙as˙D2DSHR is "true" and D2D˙Device˙connects˙to˙D2D-Relay is "true", and Number˙of˙users˙that˙device˙shares is "high", then the Resulting Case is "left". *In this case, the densification of network and offloading will be executed.*

4.7 IMPLEMENTATION OF REINFORCEMENT LEARNING AT THE DAI FRAMEWORK

In this section, we outline how reinforcement learning can be adopted in a DAI framework related to QoS and QoE. Specifically, we discuss how the Beliefs of the BDIx agent can be enhanced with a neural network that evaluates the executed actions of the BDIx agent, in terms of QoS and QoE threshold values set by the operator and the user respectively. The neural network can be implemented as a back-propagation neural network (BPNN) as in Ref. [20], trained by the UE user data rate usage every 30 minutes (set empirically). At the same time, the agent, via APIs, can request from the operator the QoS data rate value that the UE user must have. Therefore, the "UE requested data rate (QoE)", the "Operator offered data rate based on user packet (QoS)" and the "current achieved data rate" can be used as a training feature set. We envision the result of the BPNN to be a negative or positive number showing how much the current data rate is not achieving the data rate required by the user or the provided by the operator. The number that represents the achievement of QoS/QoE in the framework can be calculated using the following equation:

$$\text{Result} = \text{Current˙data˙rate} - \max(\text{QoE˙data˙rate}, \text{QoS˙data˙rate}) \tag{4.1}$$

Initially, the BPNN can be initially trained by having as feature values the QoE to be equal with the QoS provided by the Operator and the current achieved data rate to be a random variable that will get values from 0 to QoS^2. For each action the BDIx agent performs, the BPNN evaluates if the result aligns with the QoE and QoS. The BPNN will return negative numbers if the QoE and QoS are not achieved, and positive numbers if the aforementioned metrics are achieved.

Specifically, the associated Desires "Achieve QoS specified by 5G requirements" and "Achieve QoE specified by User according to historical and current records" are always running with priority 100% as Intention in the BDIx agent. These Desires have a common Plan (called "Preserve Data Rate more than the Minimum Data Rate acceptable") that monitors the data rate of the UE after each execution and tackles the not achievement of QoE or QoS.

4.8 COMPARISON OF DAI FRAMEWORK WITH RELATED WORK ON D2D FRAMEWORKS

In this section, we compare the D2D frameworks shown in Section 2.5.2 with the proposed DAI framework.

Compared to the frameworks in Refs. [21–23], the proposed DAI framework is both autonomous and distributed (for more details, see Section 2.1) and implements DAI control. This makes our approach faster in decisions and capable to tackle any situation in the mobile environment independently, as it does not depend on information that are gathered from other devices (e.g., BS) for the decision-making process. However, our approach does not support content caching. With content caching integrated, our DAI framework is expected to further improve the average downloading data rates, reduce energy consumption and improve latency. Considering the aforesaid expected gains, content caching will be investigated further as a future direction and integrated as an additional feature of our DAI framework. This will be achieved in a similar approach as in Ref. [23].

Compared to Ref. [24], in our proposed DAI framework (shown in Chapter 4), the D2D device is authorized and authenticated by the BS with the use of public key cryptography and timestamp for no repudiation attacks in a security protocol that shares tokens among the D2D devices. The aforesaid is performed upon the D2D device entering the D2D network. Once authorized and authenticated, the D2D device can communicate with other D2D devices in the network. Compared to the frameworks in Refs. [25–31], in our proposed DAI framework, the devices focus in all modes of Transmission Mode Selection. Therefore, our approach implements cluster formation along with back-hauling formation with the use of ProSe for device discovery. However, the examined framework approaches focus only on one Transmission Mode, the "D2D Relay (D2DSHR)", and therefore, they focus on the relay selection and cluster formation.

Compared to the framework in Ref. [32], through the proposed DAI framework, the devices can exchange messages and collaboratively tackle a disaster because they are distributed, autonomous and independent of the BS and other nodes. More specifically, the DAI framework can provide self-organized network (SON) operations and achieve a network recovery in a case of a mobile network failure (e.g., BS failure). Additionally, our framework utilizes all Transmission Modes in the disaster recovery and not only D2DMHR.

Overall, our proposed DAI framework, compared to all other related approaches described above, differentiates in the following characteristics: (i) our proposed framework focuses on the achievement of 5G and beyond requirements by jointly tackling all the D2D challenges in one framework; (ii) is autonomous and distributed; (iii) Uses Distributed Artificial Intelligence (DAI) control; (iv) it supports self-organize network (SON) features; (v) focuses on software agents; (vi) focuses on the local environment rather than the global network picture; and (vii) supports all transmission modes.

REFERENCES

1. S. Wen, X. Zhu, Z. Lin, X. Zhang, and D. Yang, "Distributed resource management for device-to-device (D2D) communication underlay cellular networks," *IEEE International Symposium on Personal, Indoor and Mobile Radio Communications, PIMRC*, pp. 1624–1628, London, Sep. 2013.
2. D. H. Lee, K. W. Choi, W. S. Jeon, and D. G. Jeong, "Two-stage semi-distributed resource management for device-to-device communication in cellular networks," *IEEE Transactions on Wireless Communications*, vol. 13, no. 4, pp. 1908–1920, 2014.
3. D. Wu, Y. Cai, R. Q. Hu, and Y. Qian, "Dynamic distributed resource sharing for mobile D2D communications," *IEEE Transactions on Wireless Communications*, vol. 14, no. 10, pp. 5417–5429, 2015.

4. F. Librino and G. Quer, "Distributed mode and power selection for non-orthogonal D2D communications: A stochastic spproach," *IEEE Transactions on Cognitive Communications and Networking*, vol. 4, no. 2, pp. 232–243, 2018.
5. Y. Cai, H. Chen, D. Wu, W. Yang, and L. Zhou, "A distributed resource management scheme for D2D communications based on coalition formation game," *2014 IEEE International Conference on Communications Workshops (ICC)*, pp. 355–359, Sydney, Jun. 2014.
6. H.-H. Nguyen, M. Hasegawa, and W. Hwang, "Distributed resource allocation for D2D communications underlay cellular networks," *IEEE Communications Letters*, vol. 20, no. 5, pp. 942–945, 2016.
7. R. Yin, G. Yu, C. Zhong, and Z. Zhang, "Distributed resource allocation for D2D communication underlaying cellular networks," *2013 IEEE International Conference on Communications Workshops, ICC 2013*, pp. 138–143, Houston, TX, Jun. 2013.
8. G. Fodor and N. Reider, "A distributed power control scheme for cellular network assisted D2D communications," in *2011 IEEE Global Telecommunications Conference - GLOBECOM 2011*, pp. 1–6, IEEE, Dec. 2011.
9. I. F. Akyildiz, S. Nie, S. C. Lin, and M. Chandrasekaran, "5G roadmap: 10 key enabling technologies," *Computer Networks*, vol. 106, pp. 17–48, 2016.
10. M. Herlihy and N. Shavit, *The Art of Multiprocessor Programming*. Burlington, MA: Morgan Kaufmann, 2012.
11. I. Foster, *Designing and Building Parallel Programs: Concepts and Tools for Parallel Software Engineering*. Delhi: Addison-Wesley, 1995.
12. G. O'Hare and N. Jennings, *Foundations of Distributed Artificial Intelligence*. Hoboken, NJ: Wiley, 1996.
13. R. Kruse, "Fuzzy neural network," *Scholarpedia*, vol. 3, no. 11, p. 6043, 2008.
14. H. Zhang, L. Venturino, N. Prasad, P. Li, S. Rangarajan, and X. Wang, "Weighted sum-rate maximization in multi-cell networks via coordinated scheduling and discrete power control," *IEEE Journal on Selected Areas in Communications*, vol. 29, pp. 1214–1224, Jun. 2011.
15. Qualcomm, "LTE direct is creating a digital 6th sense through always-on proximal discovery services," *Qualcomm*, September 2015.
16. "WiFi Direct: The worldwide network of companies that brings you Wi-Fi." [Online]. Available at: https://www.wi-fi.org/, Accessed on: 2020-09-19.
17. "What is LTE Direct: Device to Device Communication — 3GLTEInfo." [Online]. Available at: http://www.3glteinfo.com/lte-direct/, Accessed on: 2021-01-25.
18. Qualcomm, "LTE Direct; The Case for Device-to-Device Proximate Discovery," *Qualcomm*, 2013.
19. "Wi-Fi Direct Specification," *WiFi Alliance*, 2020.
20. A. T. Goh, "Back-propagation neural networks for modeling complex systems," *Artificial Intelligence in Engineering*, vol. 9, pp. 143–151, Jan. 1995.
21. H. J. Kang and C. G. Kang, "Mobile device-to-device (D2D) content delivery networking: A design and optimization framework," *Journal of Communications and Networks*, vol. 16, no. 5, pp. 568–577, 2014.
22. B. Bai, L. Wang, Z. Han, W. Chen, and T. Svensson, "Caching based socially-aware D2D communications in wireless content delivery networks: A hypergraph framework," *IEEE Wireless Communications*, vol. 23, no. 4, pp. 74–81, 2016.

23. N. Kumar, S. N. Swain, and C. Siva Ram Murthy, "A novel distributed Q-learning based resource reservation framework for facilitating D2D content access requests in LTE-A networks," *IEEE Transactions on Network and Service Management*, vol. 15, no. 2, pp. 718–731, 2018.
24. A. Ometov, P. Masek, J. Urama, J. Hosek, S. Andreev, and Y. Koucheryavy, "Implementing secure network-assisted D2D framework in live 3GPP LTE deployment," *2016 IEEE International Conference on Communications Workshops, ICC 2016*, pp. 749–754, Kuala Lumpur, May 2016.
25. L. Pu, X. Chen, J. Xu, and X. Fu, "D2D fogging: An energy-efficient and incentive-aware task offloading framework via network-assisted D2D collaboration," *IEEE Journal on Selected Areas in Communications*, vol. 34, no. 12, pp. 3887–39014, 2016.
26. S. Yu, R. Langar, and X. Wang, "A D2D-multicast based computation offloading framework for interactive applications," *2016 IEEE Global Communications Conference, GLOBECOM 2016 - Proceedings*, pp. 16–21, Washington, DC, Dec. 2016.
27. S. Doumiati, H. Artail, and K. Kabalan, "A framework for clustering LTE devices for implementing group D2D communication and multicast capability," *2017 8th International Conference on Information and Communication Systems, ICICS 2017*, pp. 216–221, Irbid, Apr. 2017.
28. A. Asadi, V. Mancuso, and R. Gupta, "Dore: An experimental framework to enable outband D2D relay in cellular networks," *arXiv*, vol. 25, no. 5, pp. 2930–2943, 2017.
29. A. A. Ateya, A. Muthanna, and A. Koucheryavy, "5G framework based on multi-level edge computing with D2D enabled communication," *International Conference on Advanced Communication Technology, ICACT*, vol. 2018, pp. 507–512, PyeongChang, Feb. 2018.
30. S. Doumiati, M. Assaad, and H. A. Artail, "A framework of topological interference management and clustering for D2D networks," *IEEE Transactions on Communications*, vol. 67, pp. 7856–7871, Nov. 2019.
31. X. Yi, L. Pan, Y. Jin, F. Liu, and M. Chen, "EDirect: Energy-efficient D2D-assisted relaying framework for cellular signaling reduction," *IEEE/ACM Transactions on Networking*, vol. 28, no. 2, pp. 860–873, 2020.
32. A. Thomas and G. Raja, "FINDER: A D2D based critical communications framework for disaster management in 5G," *Peer-to-Peer Networking and Applications*, vol. 12, no. 4, pp. 912–923, 2019.

Notes

[1]More specifically, DNN can be utilized for the estimation of the reused/non reused frequency. Furthermore, with the use of a DNN that is pre-trained with Channel Quality Indicator (CQI) information, distance, frequency, interference, and signal quality, the DNN will give as a result the desired frequency, which can be an existing one in the mobile network (inband underlay) or a newly assigned frequency (inband overlay) in terms of frequency mode.
[2]The ProSe as a search function allows users in proximity to discover each other
[3]LTE Direct utilizes the licensed spectrum of the intercommunicating D2D devices (underlay/overlay).

5 Example Plans of the DAI Framework/BDIx Agents to Satisfy D2D Challenges

This chapter provides example Plans on how the DAI framework and BDIx agents can be realized to satisfy specific D2D challenges. The D2D challenges that are tackled are: (i) device discovery; (ii) mode selection; (iii) mmWave communication; (vi) radio resource allocation and interference management; (v) cell densification and offloading; (vi) power control; (vii) security; (viii) handover management; and (ix) QoS & QoE. Each D2D challenge is associated with a specific Desire and a specific plan that the BDIx agent will execute towards its achievement. Additionally, for the execution of some plans, different approaches are proposed, and a specific plan is selected thereof.

In the examples provided below, an already implemented D2D communication network with its D2D devices is assumed. Additionally, the D2DMHRs and D2DSHRs devices in the network share information via LTE ProSe services (as shown in Section 5.1). For clarity, we will also use the term D2D-Relay to represent both the direct D2DSHR and the D2DMHR (i.e. D2DSHR/D2DMHR) cases.

5.1 DEVICE DISCOVERY

An important task of the DAI framework is to tackle the device discovery challenge and allow a BDIx agent to identify and locate other BDIx agents (running on D2D-R) across the same cell. To achieve this, the DAI framework utilizes specific mobile technologies and specific channels in the mobile frequency signal spectrum.

The information that can be shared between D2D devices (i.e., D2DSHRs and D2DMHRs) during device discovery, is as follows: (i) location coordinates; (ii) inband and/or outband frequency bands (channel information, mobile frequency); (iii) value of the adopted metric (e.g., WDR); iv) data rate achieved; v) transmission mode used; (vi) number of D2D Clients supported (for D2DSHR); (vii) IP and subnet of IPs assigned to D2D Clients supported (for D2DSHR); (viii) data rate achieved with other D2D devices (if any) and the BS; and (ix) channel quality indicator (CQI) measured from proximate D2D devices (D2D-Relays) and the BS/AP.

Towards this end, according to Table 4.4 and Figure 4.1, to achieve the device discovery challenge, the Desire "Identify the surrounding D2DSHRs and D2DMHRs" will get 100% priority (as shown in Section 4.4). The plan associated with this Desire is named "Gather Information from the surrounding D2DSHRs, and D2DMHRs". The following approaches can be exploited for executing the plan:

DOI: 10.1201/9781003469209-5

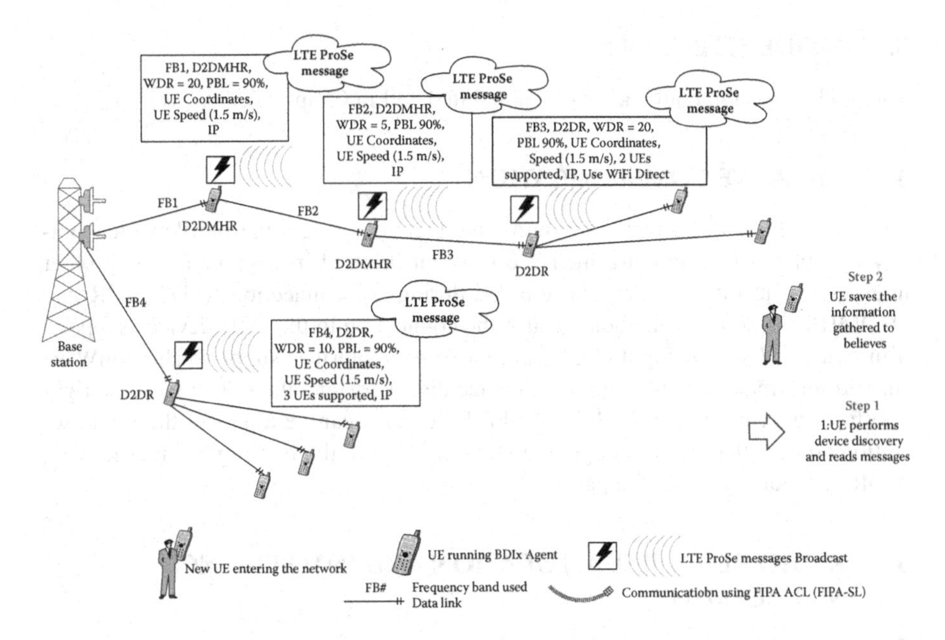

Figure 5.1 Device discovery example.

Approach 1: The common pilot channel (CPICH) or physical broadcast channel (PBCH) can be utilized for broadcasting the required information periodically without expectation of a response from the receiver (as shown in Refs. [1–3]).
Approach 2: Discovery signals can be sent or received during the specific discovery time (as shown in Ref. [4]).
Approach 3: A BDIx agent broadcast a request either through a predefined cellular or Wi-Fi[1] channel. The responding BDIx agent will use the same type (cellular or Wi-Fi) of a predefined channel to provide the requested information.
Approach 4: By using IP and by broadcasting a message at the network of the telecom operator that shares the cell connected to it, the UEs in the same cell will reply with a message that has the requested information.
Approach 5: Use LTE proximity services (LTE ProSe). The LTE ProSe is an existing solution for the device discovery that BDIx agents of the DAI framework can already utilize; it has a simple one-step plan (algorithm) to tackle device discovery.

An example realization of device discovery, using Approach 5, is provided in Figure 5.1. The following steps are executed in the example:

Step 1: The D2D candidate device executes device discovery with LTE ProSe.
Step 2: The D2D candidate device reads and saves the information from LTE ProSe messages.

5.2 MODE SELECTION

The mode selection plan will be discussed in detail in Chapter 6.

5.3 MMWAVE COMMUNICATION

For DAI framework to tackle the mmWave D2D challenge, the mmWave technology is supported as an extra interface that can be used in conjunction with other interfaces. The mmWave can be utilized during the connection to D2DSHR and D2DMHR of the D2D candidate and if the connection to the D2D devices supports it in terms of Line Of Sight (LOS) and mmWaves interface support. The mmWave support information can be sent over device discovery, and the selection of the right interface (e.g., mmWave, Mobile, Wi-Fi, Bluetooth) can be a part of the mode selection process that have as target the maximization of the investigated metric (e.g., WDR, data rate) towards the gateway.

5.4 RADIO RESOURCE ALLOCATION AND INTERFERENCE MANAGEMENT

The DAI framework aimed to tackle the radio resource allocation must also tackle the interference management jointly as they are directly associated. Interference management aims to find the best frequency bands (FBs) that causes the least interference in the network. Radio resource allocation, on the other hand, aims to identify the best way to share the found FBs within the mobile network, in such a way that interference is reduced and the data rate of the UEs is increased. Hence, BDIx agents need to tackle both D2D challenges at once with one plan.

To achieve the radio resource allocation and interference management challenge, the Desire called "Calculate Best Frequency Band with the least Interference" will get 99% priority value (see Table 4.4, Figure 4.1 and Section 4.4). The plan associated with this Desire is named "Calculate Best Frequency Band with the least Interference". For executing this plan, the following approaches can be exploited:

Approach 1: The OFDMA scheme underlay approach, in which the BS has a pool of orthogonal frequency bands to share among the D2D devices connected to the BS. With this approach, the utilization of reused or free FBs (underlay type of spectrum utilization) among the D2D device is done as follows: (i) When the device discovery task is completed, the broadcasted information that is collected by the D2D candidate from the D2DSHRs and D2DMHRs, includes the FBs used by the D2D-Relay along with their coordinates and their D2D Clients; (ii) Then the D2D device can identify which reused FB to utilize with the least interference according to a minimum acceptable interface predefined threshold; (iii) Finally, the FB with the least interference is evaluated with a use of deep neural network (Feed Forward Network); (iv) The Deep Neural Network (DNN) is trained using simulation/live data. The features that we can be used as inputs are: (a) coordinates of D2D device; (b) coordinates of the candidate cevice; (c) CQI; (d) transmission power; and (e) FB used. (v) The result of DNN, using a specific FB, is the interference of the examined devices, and if the result is above the interference threshold, then the examined FB can be used by the D2D candidate device. An important note about the approach is that when the D2D candidate establishes a connection with the BS, the FB given by the BS is not lost. The assigned FB, along with the calculated FB (from the previous step) are saved in the Beliefs as an alternative band to be used in the process of the transmission mode selection process or when interference increases.

Approach 2: The sharing information with BS underlay approach, in which the BDIx agent of a candidate D2D device tries to identify the best FB to be used for communication among the D2D devices and the BS. The BS also has the DAI framework BDIx agent installed as a component and acts as a D2DSHR. So, the BS can share the same information as other D2DSHRs and D2DMHRs in the device discovery process. With this approach, the utilization of FBs (underlay type of spectrum utilization) among the D2D device is done as follows: (i) After the device discovery phase, the D2D candidate device can gather all the FBs, coordinates, and types[2] of communication protocol (inband, outband) used in the communication network from D2DSHRs and D2DMHRs and their D2D Clients. (ii) Subsequently, the D2D candidate, after selecting the D2D-Relay with the maximum distance it requests its FB and coordinates (inband - LTE direct). In the case of single hop D2D-Relay the D2D candidate requests the same information from the its furthest D2D Client (outband and Wi-Fi Direct); and (iii) Afterwards, the D2D candidate calculates the interference among the two D2D devices in two cases:

- In the case of inband, the D2D candidate calculates the interference, and if lower than a threshold it reuses the same FB. Otherwise, it reduces the transmission power according to its QoS and QoE requirements to reduce interference and thus remain lower than the threshold (note: interference can be calculated using Shannon theorem). Successively, if the steps above did not calculate the reused FB, then the D2D candidate device can use the FB provided by the BS.
- In the case of outband, the D2D candidate calculates the best reused outband frequency channel by following the same procedure as with the inband case. If a reused frequency channel is not found, then the D2D Client accepts the frequency channel that the D2DSHR provides.

Approach 3: The radio resource allocation is done in terms of interference management using the communication advantage that the BDIx agents have regarding accessing all devices in the network. This approach can be based on a coalition game (as shown in Refs. [5–7]) in game theory. In this case, all the D2D devices participate in a bazaar-style "give and take" in terms of resource blocks and transmission power with reused frequencies by having a utility function to target the maximization of Sum Rate. An example of distributed auction-based game theory in D2D communication is shown in Refs. [8–10].

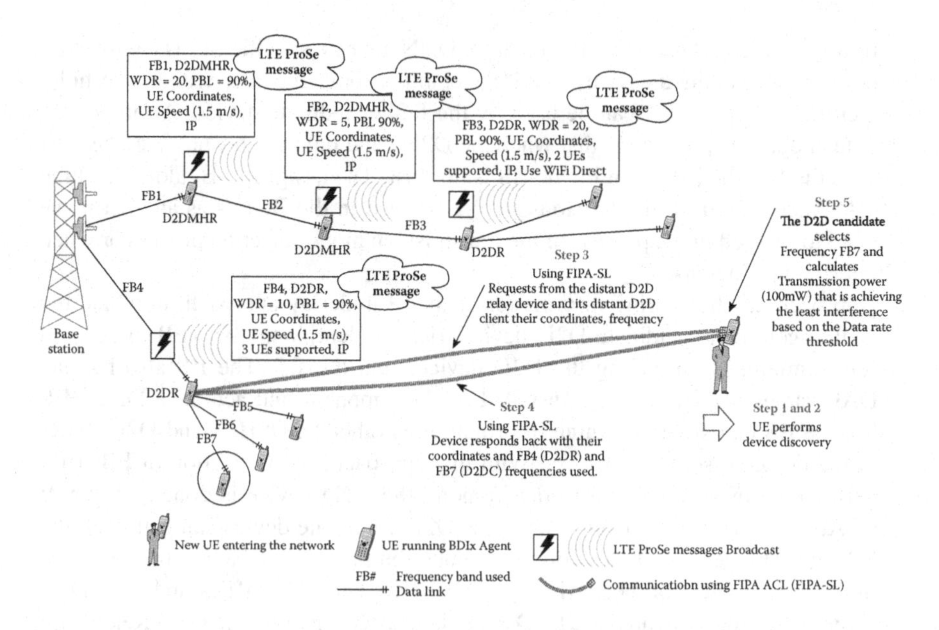

Figure 5.2 Radio resource allocation & interference management example.

An example realization of "Radio Resource Allocation and Interference management", using Approach 2, is provided in Figure 5.2. The following steps are executed in the example:

Step 1: The D2D candidate device executes device discovery with LTE ProSe.
Step 2: The D2D candidate device reads and saves the information from LTE ProSe messages.
Step 3: The D2D candidate device requests from the most distance D2DSHR (which is with FB4) its distant D2D Client with their coordinates and FBs.
Step 4: The D2DSHR (with FB4) replies with is coordinates and FBs.
Step 5: The D2D candidate selects FB and calculates the transmission power (TP) that guarantees the Data Rate requested with the least interference caused.

5.5 CELL DENSIFICATION AND OFFLOADING

For the DAI framework to tackle the "Cell Densification and Offloading" challenge, a plan must be implemented that will focus on D2DSHR offloading when the cell is dense in bandwidth utilization or the number of users it can support. The issues mentioned above are tackled by moving D2D Clients from a dense D2DSHR to another D2DSHR (called receiver D2DSHR) that is not as dense, which is near the D2D Client (e.g., less than 200 m). Additionally, if there are excess D2D Clients that the receiver D2DSHR can handle after the movement execution, these excesses of

D2D Clients can connect directly to the BS. Thus, there are two cases of Desires associated with this D2D challenge. These cases are the following:

- The case of Dense D2DSHR Node: According to Table 4.4 and Figure 4.1, to achieve the "Cell Densification and Offloading" challenge, the Desire "Number of D2D Clients that the D2D device serves as D2DSHR is acceptable" will get 99% priority (as shown in Section 4.4).
- The case of high bandwidth utilization at the D2DSHR Node and offloading: According to Table 4.4 and Figure 4.1, to achieve the "Cell Densification and Offloading" challenge, the Desire "Bandwidth consumed by users that the D2D device serves as D2DSHR is acceptable" will get 99% priority (as shown in Section 4.4).

The plan that both Desires will execute is the same, and it is called "Move a percentage of D2D Clients to other D2DSHR". In this plan, each BDIx agent that acts as a D2DSHR in the D2D communication network has a constrained value for the maximum number of clients for the cluster head and a bandwidth percentage threshold. Thus, if in a D2DSHR the maximum number of D2D Clients is reached, then the following steps are executed: (i) this D2DSHR will communicate with the nearest D2DSHR offering to move a number of D2D Clients to it; (ii) if the nearest D2DSHR accepts the proposal, then the dense D2DSHR sends the D2D Clients to connect to the contacted D2DSHR; (iii) if the nearest D2DSHR did not accept the request proposal, the dense D2DSHR sends the excess clients to the BS. So, in the case that the nearest D2DSHR agreed, the dense D2DSHR sends a number of D2D Clients (offloading) to the nearest D2DSHR by informing them what the next hop will be. The D2D Clients assignment procedure differs according to the type of frequency that the D2DSHR is sharing. The cases are the following:

- If the contacted D2DSHR is inband D2D, the D2D Client agrees on the frequency channel to be used with the contacted D2DSHR using radio resource allocation and interference management approaches.
- If the contacted D2DSHR mode is outband D2D, the D2D Client agrees on the specific Wi-Fi channel to be used with the contacted D2DSHR.

An example realization of "Cell Densification and Offloading", is provided in Figure 5.3. The following steps are executed in the example:

Step 1: The D2D candidate device executes device discovery with LTE ProSe.
Step 2: The D2D candidate device reads and saves the information from LTE ProSe messages.
Step 3: The D2D candidate device performs Mode Selection and selects D2D Client mode to D2DSHR (with FB 3).
Step 4: The D2D candidate device requests to connect to D2DSHR (with FB 3).
Step 5: The D2DSHR (with FB 3) is full in terms of the number of D2D Clients connected to it (i.e., it has reached the 200 D2D Clients threshold). So, it should

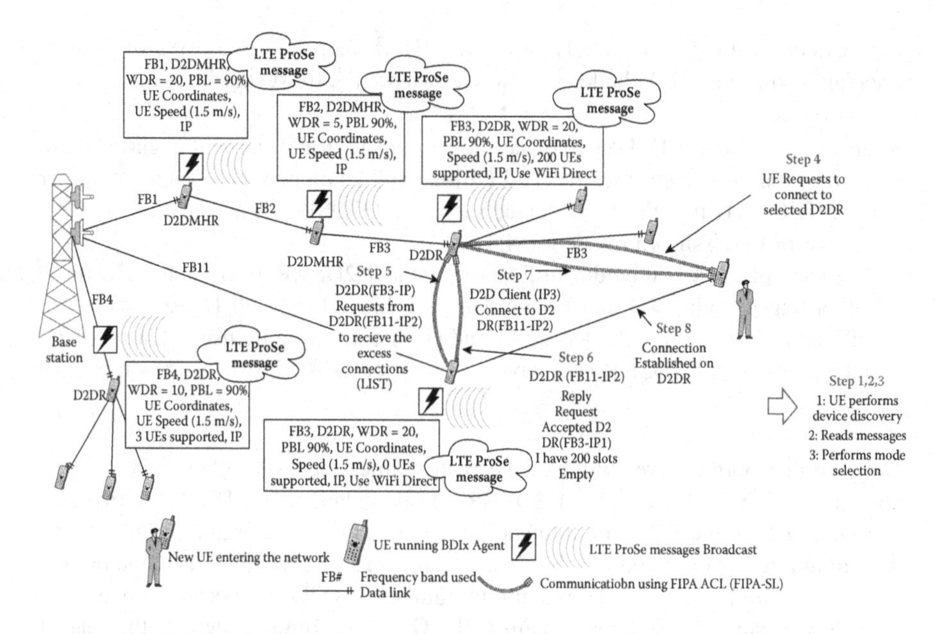

Figure 5.3 Cell densification and offloading example.

move the excess D2D Client to another D2DSHR. Thus, the D2DSHR intercommunicates with the nearest D2DSHRs that are satisfying the requirements of distance and data rate (along with other thresholds) and reallocates the D2D Client to the most "profitable" D2DSHR[3] in terms of a metric (e.g., Sum Rate).
Step 6: The D2DSHR (with FB 11) replies to the request of D2DSHR (with FB 3) that it can accept the new D2D device.
Step 7: The D2DSHR (with FB 3) instructs the D2D Client to connect to the D2DSHR (with FB 11).
Step 8: The D2D Client connects to the D2DSHR (with FB 11).

5.6 POWER CONTROL

Power Control (P-C) during D2D communication is essential for achieving battery reservation and extend the battery life of the device. For DAI framework to tackle the "Power Control" D2D challenge, a plan must be implemented that will focus on energy preservation. This plan will run after the system boot and be always associated with an Intention under the active pursuit of the BDIx agent with priority value 100%.

To achieve the power control D2D challenge, the Desire called "Battery Power reservation at D2D device" will get 100% priority value (see Table 4.4, Figure 4.1 and Section 4.4). The plan associated with this Desire is named "Battery Power Reserve". The Intention/Desire with the associated Plan will execute the task of reducing power consumption when the battery level value reaches the threshold of

"Minimum Battery Level Threshold exceeded" and an event is raised. For executing this plan, the following approaches can be exploited:

- Approach 1: This approach targets power consumption by reducing Transmission Power (TP). The BDIx agent can reduce transmission power, however at the same time must consider related thresholds that guarantee the data rate and interference requirements (as shown in Appendix B and Section B.1).
- Approach 2: This approach targets the increase of stored power through wireless power transfer [11]. In this approach, the BDIx agent can use other/unused interfaces to start charging the D2D device battery. In order to achieve wireless power transfer, a wireless power transfer beacon (source) must exist in the BS (using a predefined frequency for power transfer) that transmits power and the associated electronics in the UE.
- Approach 3: This approach targets power consumption by the reduction of CPU and memory utilization by monitoring user applications. In this approach, the BDIx agent can propose to the user to stop some applications that are currently running on the device and not being used, or even take the permission from the user to automatically stop applications running in the background consuming a lot of battery power. In particular, the identification of the demanding user applications can be made using device statistics gathered from the log or with the data collected by the BDIx agent from the users' activities regularly 24/7. This data includes application, power consumption, and timestamp.

An example realization of "Power Control", using Approach 1, is provided in Figure 5.4. The following steps are executed in the example when the threshold of "Minimum Battery Level Threshold exceeded" is raised:

Step 1: The D2D Client device calculates the transmission power reduction.
Step 2: The D2D Client device informs the D2DSHR of the TP reduction it has decided.
Step 3: The D2DSHR responds "OK" to its D2D Client.

5.7 SECURITY

Security is essential for the successful implementation and the realization of the communication among the D2D devices during D2D communication. For DAI framework to tackle the "Security" D2D challenge, a plan associated with a Desire must be implemented that will focus on monitoring and tackling any incidents related to the security aspects of the device and the security aspects of the utilized network all the time. Therefore, for this plan, there are some tasks (e.g., check link utilization) that need to be implemented regarding communication, prevention, identification of attack, and device security. This plan will run after the system boot and be always associated with an Intention under the active pursuit of the BDIx agent with priority value 100%.

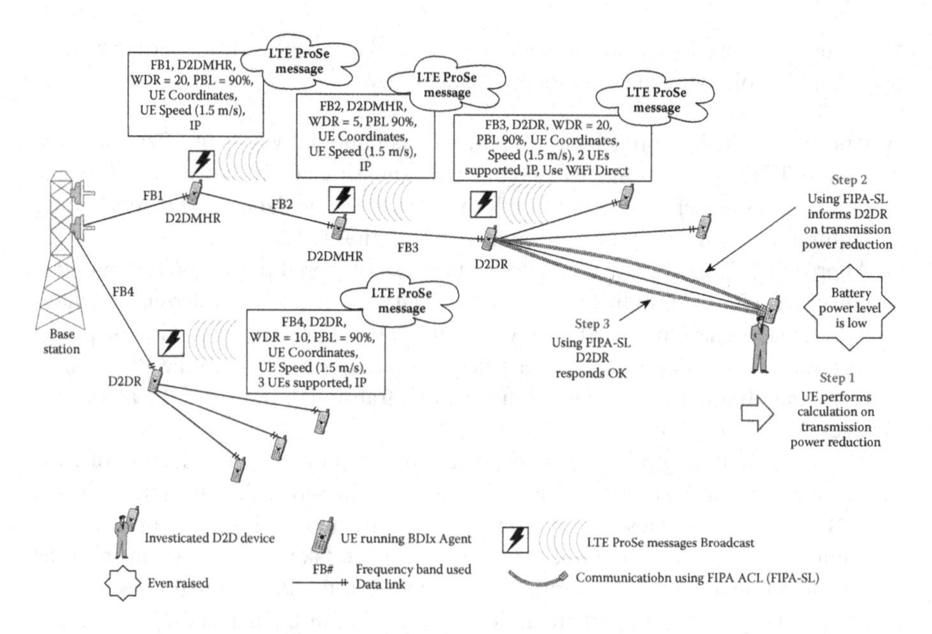

Figure 5.4 Power reservation example.

To achieve the security D2D challenge, the Desire called "Security Monitoring at D2D device" will get 100% priority value (see Table 4.4, Figure 4.1 and Section 4.4). The plan associated with this Desire is named "Security Monitoring and Intrusion Detection". In the case of an event related to security, the security plan has a logic to identify the relating event by monitoring the beliefs that are related to security in BDIx agent (e.g., monitoring the "Security Breach" boolean value in beliefs). The following approaches can be implemented in the plan, either as a protocol or device O/S:

Approach 1: This approach executes security checks related to the Transmission Mode selected by the D2D device. Thus, the following cases are handled:

D2DSHR: The D2D device has counters and AI/ML logic to evaluate the packets if they are benign or adjective. For example, the D2D device could use binary logistic regression [12] to identify if the packets sent from a device to it are benign. Also, if a D2D device is attacking a specific D2D device, the latter can inform the D2DSHR device to blacklist its former client device.

D2D Client: The D2D device evaluates the successful flow of packets using confirmation messages. For example, it evaluates the packets arrived and executes specific check-ins in terms of security (e.g., timestamp, source). If an issue arises with the connection to D2DSHR, the D2D Client should re-execute mode selection, excluding the existing D2DSHR.

Approach 2: In this approach, a security protocol for D2D is implemented in the communication among D2D devices. The utilization and enforcement of the security protocol among the D2D devices is done as follows: (i) The secure protocol will force the D2D devices that want to communicate among themselves to have a digital signature assigned from the telecom operator hardcoded for the specific D2D device; (ii) All the packets sent for intercommunication among the D2D devices should be signed and encrypted by the destination public key[4] using public-key cryptography along with timestamp token signed from BS and generated from the BS in the phase of D2D device authentication[5] and authorization for the reason of no repudiation and personalization of the D2D device; (iii) The recipient D2D device will use its signature private key to decrypt. In this approach, blockchain technology, which is decentralized, can be adapted to be used for token generation and sharing in the same manner as in the BS.

Approach 3: This approach targets the D2D device O/S, to secure the BDIx agent as software at a device that can be rooted. Therefore, the primary concern is where the BDIx agent will reside with the significant concern to be secure; i.e., no one with rooting knowledge and any device with routing capabilities to be able to access the BDIx agent and change its beliefs. To secure the agent, the BDIx agent must reside in a container-based (docker container) environment under the O/S of the device that will not be accessible even with the docker commands (isolated). This container can have access to full device capabilities (hardware and software libraries). The aforesaid container-based approach will guarantee the security of the BDIx agent. Also, the agent can be upgradeable from the telecom operator. This can be done with the use of an API and the authorization parameters of the telecom operator that are hardcoded in the device's SIM.

An example realization of how "Security" D2D challenge is satisfied with the DAI framework, is provided in Figure 5.5. The following steps are executed in the example:

Step 1: The D2D Client device executes denial Of service at the D2DSHR (with FB3).

Step 2: The D2DSHR (with FB3) requests from the D2DSHR (with FB11) to disconnect the attacking D2D Client.

Step 3: The D2DSHR (with FB11) accepts the request from the D2DSHR (with FB11) to disconnect the attacking D2D Client.

Step 4: The D2DSHR (with FB11) informs the attacker D2D Client to stop.

Step 5: The D2D Client did not respond to the request (waiting for response timeout).

Step 6: The D2DSHR (with FB11) disconnects the attacking D2D Client.

5.8 HANDOVER MANAGEMENT

The handover D2D challenge has a partial implementation rule. Specifically, it is fully implemented by tackling either one or both (depending on the event raised) of the "Mode Selection" and "Cell Densification and Offloading" challenges (see Figure 4.1). Thus, when an event is raised, the only thing that the Handover plan has

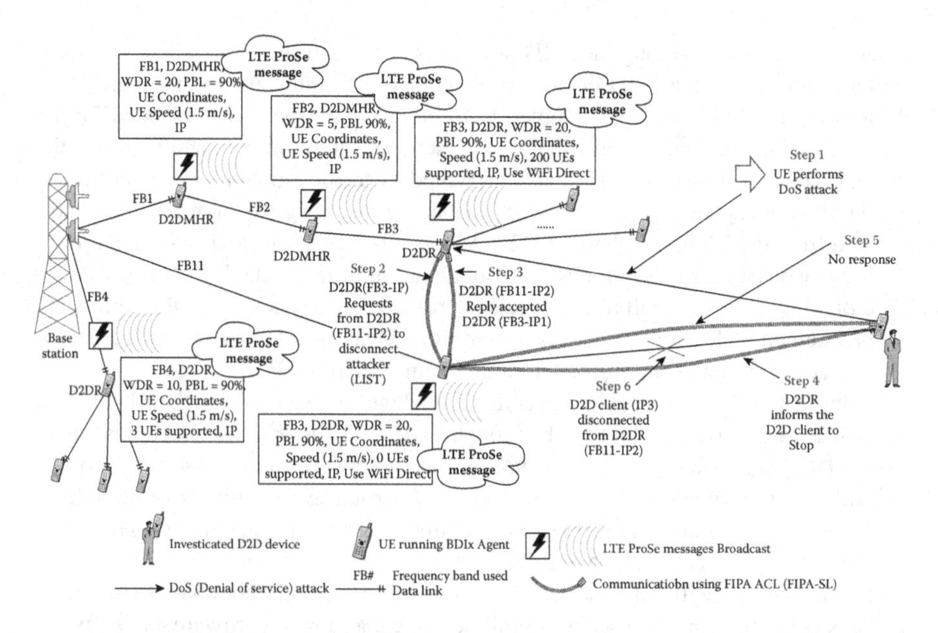

Figure 5.5 Security example.

to do is to monitor (from the Belief values) the completion status of the above challenges. If either one or both are completed, then handover is set as completed as well.

5.9 QOS AND QOE

The "QoS and QoE" D2D challenge is essential during D2D communication for achieving the targets that the user[6] and the telecom operator[7] set for the BDIx agent as threshold values (e.g., minimum data rate, maximum latency). For DAI framework to tackle the "QoS and QoE" D2D challenge, a plan must be implemented that will aim to increase data rate and reduce latency. This plan will run after the D2D device joins in a D2D communication Network and be always associated with an Intention under the active pursuit of the BDIx agent with priority value 100%. Note that there are cases in the DAI framework where the QoS and QoE are tackled with different Desires. For achieving the "QoS and QoE" challenge, the following two cases can be used:

- Case 1: The Desires "Achieve QoS specified by 5G requirements" and "Achieve QoE specified by User according to historical and current records"[8] will get 100% priority value (see Table 4.4, Figure 4.1 and Section 4.4). The plan associated with the aforesaid Desires is called "Preserve Data Rate more than the Minimum Data Rate acceptable" and monitors the data rate of the UE after each execution. This plan is initiated upon the D2D device entering the D2D network and then always be under the active pursuit of the agent. When an event is raised, this plan identifies its type by monitoring the beliefs related to the associated "QoS and QoE" events. The plan can be implemented using the following approaches:

- Approach 1: "Mode selection" Desire increases priority value to 98%.
- Approach 2: Increase the transmission power, if possible. Therefore, the BDIx agent can change the transmission power to achieve the QoS required data rate.

It is worth mentioning here that in case "Power Save" Desire becomes Intention, the above Desires will cease execution and set their priority value to 0%. The reason is that Desires relates to Power reservation and considered more important for the user than the Desires related to QoE & QoS.

- Case 2: The Desire called "The latency (round time/ultra-reliable low latency communication) of accessing gateway or any other D2D device is acceptable" will get 100% priority value (see Table 4.4, Figure 4.1 and Section 4.4). The Desire can be achieved with the "Preserve the latency low" plan that is targeting low latency by achieving routing. The plan can be implemented using the following approaches:
  - Approach 1: The D2DSHRs and D2DMHRs use "Device Discovery" to learn about the D2D communication structure, by reading the broadcast messages that have the D2D-Relay coordinates. With this approach, the QoS and QoE achievement is done as follows: (i) The D2D device, by using a threshold distance metric, learns: (a) about the nearby D2DSHR and D2DMHR nodes that can act as next-hop routes;[9] and (b) about the D2D-Relays that can be exploited for multi-hop path construction; (ii) Using the Dijkstra[10] or Bellman-Ford algorithm,[11] each D2D-Relay calculate the best paths towards all other D2D-Relays and constructs its routing table; (iii) The D2D-Relay forwards packets to the destination D2D device according to the best route in its routing table.
  - Approach 2: Utilize SOM (self-organizing map) unsupervised learning to train the artificial neural network (ANN) using test packets. Test Packets will include as features to the ANN the following: destination, coordinates, data rate, round time and next hop. The whole process must be executed at the beginning of the network formation, and then it will follow the resulting next-hop location that the SOM will give based on the classification result that is trained.

An example realization of "QoS & QoE" D2D challenge using the DAI framework, is provided in Figure 5.6. The following steps are executed in the example, when the data rate value reduces more than the threshold of "Minimum Data Rate Acceptable":

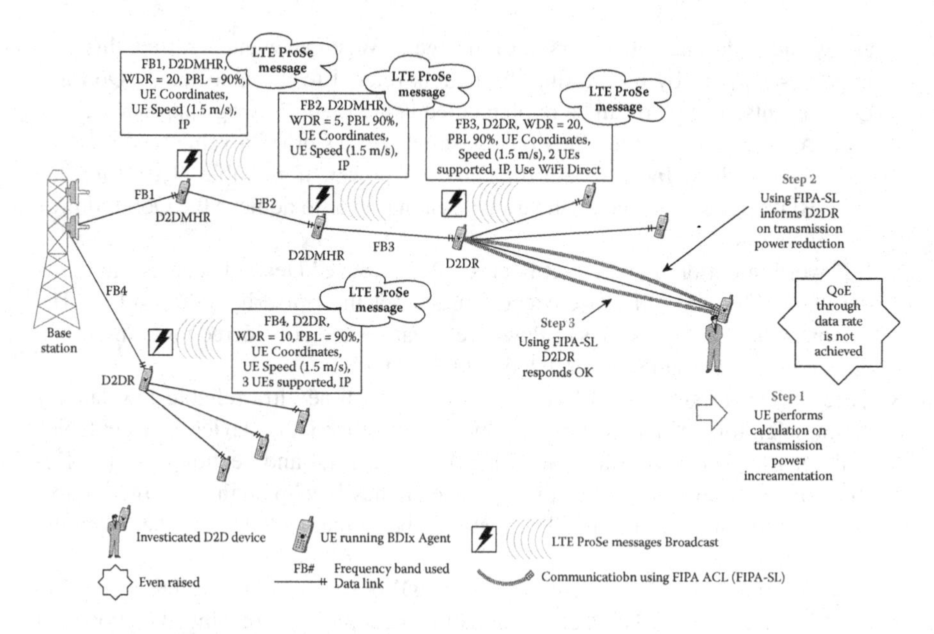

Figure 5.6 QoS & QoE example.

Step 1: The D2D Client device calculates the Transmission Power reduction.
Step 2: The D2D Client device informs the D2DSHR, that is connected to it, the TP reduction that it decided.
Step 3: The D2DSHR responds "OK" to its D2D Client.

REFERENCES

1. J. Feng, Device-to-device communications in LTE-advanced network. Theses, Télécom Bretagne, Université de Bretagne-Sud, Dec. 2013.
2. A. Gotsis, K. Maliatsos, P. Vasileiou, S. Stefanatos, M. Poulakis, and A. Alexiou, "Experimenting with flexible D2D communications in current and future LTE networks: A D2D radio technology primer & software modem implementation," *Wireless Innovation Forum European Conference on Communication Technologies and Software Defined Radio*, vol. 24, pp. 38–61, Mar. 2017.
3. H. Tang, Z. Ding, and B. C. Levy, "Enabling D2D communications through neighbor discovery in LTE cellular networks," *IEEE Transactions on Signal Processing*, vol. 62, no. 19, pp. 5157–5170, 2014.
4. O. Hayat, R. Ngah, S. Z. Mohd Hashim, M. H. Dahri, R. Firsandaya Malik, and Y. Rahayu, "Device discovery in D2D communication: A survey," *IEEE Access*, vol. 7, pp. 131114–131134, 2019.
5. E. Elkind and J. Rothe, *Cooperative Game Theory*. Berlin, Heidelberg: Springer, 2016.
6. Y. Cai, H. Chen, D. Wu, W. Yang, and L. Zhou, "A distributed resource management scheme for D2D communications based on coalition formation game," *2014 IEEE International Conference on Communications Workshops (ICC)*, pp. 355–359, Sydney, Jun. 2014.

7. C. Gao, Y. Li, Y. Zhao, and S. Chen, "A two-level game theory approach for joint relay selection and resource allocation in network coding assisted D2D communications," *IEEE Transactions on Mobile Computing*, vol. 16, no. 10, pp. 2697–2711, 2017.
8. A. Antonopoulos, E. Kartsakli, and C. Verikoukis, "Game theoretic D2D content dissemination in 4G cellular networks," *IEEE Communications Magazine*, vol. 52, no. 6, pp. 125–132, 2014.
9. S. M. Kazmi, N. H. Tran, W. Saad, Z. Han, T. M. Ho, T. Z. Oo, and C. S. Hong, "Mode selection and resource allocation in device-to-device communications: A matching game approach," *IEEE Transactions on Mobile Computing*, vol. 16, no. 11, pp. 3126–3141, 2017.
10. J. Huang, Y. Yin, Y. Sun, Y. Zhao, C. C. Xing, and Q. Duan, "Game theoretic resource allocation for multicell D2D communications with incomplete information," *IEEE International Conference on Communications*, pp. 3039–3044, London, Sep. 2015.
11. K. R. Sreelakshmy and L. Jacob, "Simultaneous wireless information and power transfer in heterogeneous cellular networks with underlay D2D communication," *Wireless Networks*, vol. 26, pp. 3315–3330, Jul. 2020.
12. C. Ioannou and V. Vassiliou, "An intrusion detection system for constrained WSN and IoT nodes based on binary logistic regression," in *Proceedings of the 21st ACM International Conference on Modeling, Analysis and Simulation of Wireless and Mobile Systems*, New York, pp. 259–263, Association for Computing Machinery (ACM), Oct. 2018.

Notes

[1] Wi-Fi is used in case there is a short distance among the communicating D2D devices.

[2] In our case, the inband is used for links among BS, D2DSHRs, and D2DMHRs. The outband is used for the link-sharing of D2DSHR to D2D Clients.

[3] Note that if there is no D2DSHR satisfying the requirements, the D2D Client connects to the BS directly as a D2DMHR.

[4] Public key will be shared among the others via the certificate authority that issued the digital signature.

[5] The device's username is its International Mobile Subscriber Identity (IMSI), Mobile Station International Subscriber Director Number (MSISDN) and International Mobile Equipment Identity (IMEI) number, and the password is a randomly generated password when the D2D device registers to the telecom operator network (when the user buys the sim card).

[6] The QoE is associated with the Desires of the user in terms of bandwidth and data rate.

[7] The QoS is a associated with the minimum bandwidth that the D2D device must achieve. The telecom operator can guarantee this.

[8] The QoE Desire has as a target to check if there is a lack of data rate, through the examination of current loaded applications and from examining historical events and with the use of BPNN and RL as shown in the 3.4.

[9] For calculating next-hop, the threshold value of "Maximum Distance of another Node to be a Neighbour"is considered.

[10] Dijkstra is in principle a routing algorithm that requires global knowledge, thus in order to consider it in the investigation we assume that all D2D-Relay devices share over LTE ProSe and thru messages their routing information.

[11] These algorithms run on each D2D-Relay device and take into consideration weights like distance, WDR, time to access BS, etc.

6 DAI Framework for Addressing the D2D Mode Selection Challenge

This chapter introduces the distributed artificial intelligence solution (DAIS) plan, which aims to provide an illustrative example of how the DAI framework can be exploited for D2D mode selection (frequency & transmission). DAIS is a plan of the DAI framework focusing on establishing communication between D2D devices in proximity. It is considered by the BDIx agents, as the plan of execution for selecting the transmission mode that the D2D devices will operate upon when entering the D2D communication network. The implementation of the transmission mode selection in the book is such that the device also selects the frequency mode selection,[1] with a view to improve the spectral efficiency (SE)[2] and Power Consumption (PC)).[3]

Initially, we implement a simple DAIS plan and evaluate its performance in a static environment (as shown in Chapter 7). Building on the results of the initial evaluation, we enhance DAIS with additional functionality and further evaluate its performance. Additionally, we extend DAIS to handle dynamic aspects. Furthermore, to allow comparison with DAIS, we also introduce a centralized technique based on SR and its distributed version, labeled as Distributed Sum Rate (DSR), as well as other competing techniques (see Section 2.5.3), suitably enhanced if required to be applied in a D2D setting.

Note that a list of common parameters for DAIS and DSR (Sum Rate Approach) appear in Section A.1 in Appendix A.

6.1 DISTRIBUTED ARTIFICIAL INTELLIGENT SOLUTION PLAN FOR D2D TRANSMISSION MODE SELECTION IN STATIC AND DYNAMIC ENVIRONMENT

Distributed artificial intelligence solution (DAIS) is a specific plan of the DAI framework focusing on establishing communication between D2D devices in proximity. It is considered by the BDIx agents as the plan of execution (e.g., in the event of a D2D device entering the network), in order to select the transmission mode that the D2D devices will operate. This is achieved in a distributed artificial intelligence manner, considering the weighted data rate (WDR) metric and local network knowledge acquired through the exchange of LTE ProSe messages. The transmission mode that a D2D device will operate is selected in such a way that the weighted data rate (WDR) metric[4] is maximized in a localized manner. In this section, the Weighted Data rate (WDR) metric is introduced, representing the minimum data rate of the weakest link in a D2D communication path (directly connected to the BS or via

DOI: 10.1201/9781003469209-6

D2D-Relay nodes). Also, the SR is introduced, representing the summation of the Data Rate of each link in the D2D network. DAIS aims to select the transmission mode that a D2D device will operate in such a way that the WDR metric is maximized (see Section 6.1.4). In addition, the WDR and Battery Power Level (BPL) thresholds are examined. Finally, the initial implementation of the DAIS, calculations of thresholds, enhancements of DAIS, and extension of DAIS to support a dynamic environment with speed and direction are also described in this section.

6.1.1 MODE SELECTION

The "Mode Selection" (and thus the DAIS plan) is initiated when a D2D device enters the D2D communication network. In this case, the BDIx agent receives the "D2D device Enters a D2D Communication Network" event and sets the Desires' order of execution that will achieve the establishment of the "Mode Selection" challenge (see Table 4.4, Figure 4.1 and Section 4.4):

- The Desire "Find best Transmission Mode that achieves the best achievable Signal Quality, Data Rate and WDR" with the associated Plan "Distributed Artificial Intelligence Solution (DAIS)", will get 98% priority value.
- The Desire "Find the best-reused Frequency with the least Interference" with the associated Plan "Calculate Best Frequency Band with the least Interference".[5] will get 99% priority value.
- The Desire "Identify the surrounding D2DSHRs and D2DMHRs" with the associated Plan "Gather Information from the surrounding D2DSHRs, and D2DMHRs", will get 100% priority value.

6.1.2 SUM RATE AND WEIGHTED DATA RATE

One of the most common metrics for the evaluation of D2D solutions is SR. The SR is the total throughput in a network calculated as the sum of the data rates that are delivered to all UEs and D2D UEs in a network [1,2]. Variations on Sum Rate exist, such as Weighted Sum Rate in Ref. [3], which considers certain links to be of more importance and gives different weights to the links based on the mode of transmission (direct, relay, etc). We introduce a new metric called "Weighted Data Rate" (WDR). The WDR is defined at each node as the minimum data rate in the path that the UE selected. The minimum data rate of a path is the data rate of the weakest edge in the path. Our aim is, essentially, to maximize the WDR, i.e WDR = max(min(Link Data Rate) for each path. The choice for using WDR instead of sum-rate is mainly for reducing the computational load of the BDI agent. The benefits will be shown clearly in the next section. For the formulation of WDR and its calculation see Section 6.1.4.

As stated, when a D2D device enters a D2D communication network, the event "D2D device Enters a D2D Communication Network" is raised and received by the BDIx agent. Based on this event, the Fuzzy Logic plan library defines the specific Desires that should be executed as well as their order of execution (i.e., by assigning

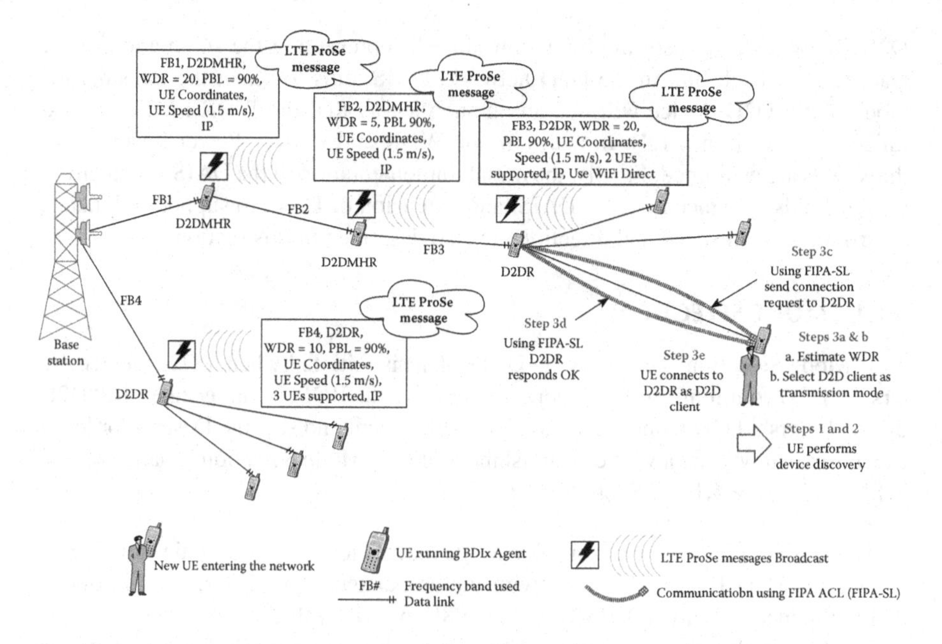

Figure 6.1 DAIS plan achieving transmission mode selection.

to them priority values starting from 100% with a decreasing step of −1) that would achieve "Transmission mode selection" for the entering D2D device. The priority value assigned to the Desires defines the steps that a BDIx agent will execute in order to achieve its target, (i.e., select the transmission mode that the D2D device will operate) upon entering the D2D network. These steps are illustrated in Figure 6.1, demonstrating a representative scenario where a D2D device selects to connect to a D2DSHR as a D2D client, and described below.

Step 1: The Desire "Device Discovery" becomes an Intention and adopted for active pursuit by the BDIx agent. This Intention is pursuit, by the agent, through the exchange of LTE ProSe messages. Through these messages, the D2D device (BDIx agent) becomes aware of all other D2D devices in its proximity including also some other information related to them (e.g., WDR of D2D device, frequency band the D2D device uses, etc.). Indicative exchanged information appears in the square boxes near the devices shown in Figure 6.1.

Step 2: Once the "Device Discovery" Intention is fulfilled the priority value of the related Desire is set to 0% while the rest of the Desires are increased by 1% (which now makes the Desire "Data Rate is acceptable" priority value equal to 100%).

Step 3: The Desire "Data Rate is acceptable" becomes an Intention and adopted for active pursuit by the BDIx agent running on the D2D device. The related Desire is associated with the "DAIS" Plan (as shown in Section 6.1.5) which goes through the following steps:

a. Estimate the WDR achieved by each D2D-Relay in its proximity and identify the one with the highest WDR (see Section 6.1.4).
b. Select "D2D Client" as Transmission mode and start the process to connect to the D2DSHR with the highest WDR.
c. Request from the D2DSHR that it wants to connect to it (the request is sent to the D2DSHR with the use of its IP address pre-acquired through ProSe messages).
d. The D2DSHR responds to the request. In the example shown in Figure 6.1, the request is accepted. Additionally, the D2DSHR adds the MAC Address of the new "D2D Client" in its "Allowed List of Devices".
e. The D2D Client connects to the D2DSHR.

Step 4: Once the DAIS plan is finalized and the Intention is achieved, the priority value of the related Desire is set to 0%. Then another Desire is selected, if any, based on the priority values set by the Fuzzy Logic rules, to become an Intention and be fulfilled by the BDIx agent.

6.1.3 WDR AND BPL THRESHOLDS

In this section, we introduce the thresholds that DAIS along with DSR are using.

6.1.3.1 WDR Threshold

The WDR Threshold refers to: (i) the minimum WDR that an existing D2D device operating as D2D-Relay must have in order for a new D2D device entering the network to connect to it; or (ii) the maximum WDR that a new D2D device entering the D2D network must have in order to replace a D2D device operating as D2D-Relay and take its role.

Thus, the WDR threshold is considered by entering a new D2D device for four purposes. The D2D device:

- Can perform a quality check of the D2DSHR, in order to connect to it as a D2D Client, by using Eq. 6.5.
- Can perform a quality check of the D2DMHR, in order to connect to it either as a D2D Multi hop Relay (DDMHR) or a D2D single hop Relay (D2DSHR), i.e. D2D-Relay, by using Eq. 6.5.
- Can replace a D2D-Relay device and take its role, if the new D2D device WDR is greater than the WDR of the existing D2D-Relay device, by using Eq. 6.8.
- Can connect to a D2D-Relay device in its proximity, and act as a D2DSHR, by using Eq. 6.7.

Based on extensive simulative evaluation, the WDR Threshold values providing the best results are: (i) 20% for the WDR for scenarios with low ($\leq$200) number of D2D devices (shown in Ref. [4]); (ii) 35% for the WDR for scenarios with large ($\leq$1000) number of D2D devices (shown in Ref. [5]). Thus, the WDR Threshold is calculated using Eq. 6.1.

$$\mathrm{WDR}_{\mathrm{Threshold}} = \begin{cases} 20\% & \text{if } N <= 200 \\ 35\% & \text{if } N > 200 \end{cases} \tag{6.1}$$

where N is number of D2D devices.

6.1.3.2 BPL Threshold

The BPL threshold determines the minimum value that the remaining battery level of a D2D device must be, in order to be able to become a D2D-Relay and accept connections from other D2D devices. The Battery Power Threshold is used by the DAIS algorithm for two purposes: (i) to limit the number of D2D devices that can be connected to a D2D-Relay device and avoid these from battery drain and (ii) avoid any QoS degradation (broken links) due to battery exhaustion. Based on extensive simulative evaluation [4, 5], the BPL threshold value providing the best results is 75% (see Eq. 6.2).

$$\mathrm{BPL}_{\mathrm{Threshold}} = 75\% \tag{6.2}$$

When the battery level of the D2D device drops below the BPL Threshold and the device acts as a D2D-R, an event is raised that increases the priority of the Desire related to the reduction of the power consumption. The aforementioned Desire becomes Intention and the Plan shown in the appendix (see Algorithm B.1 in Section B.1) is executed with a target to reduce Transmission Power with the minimum possible reduction of Spectral Efficiency (SE).

A good selection of the WDR and BPL threshold values depends highly on the identification of the number of the other D2D devices in the proximity of a D2D device, either this is entering the D2D network or executing its plan in case of a BDIx change of "Belief". To achieve this in a localized manner, the DAIS algorithm was enhanced to include sharing of the number of D2D devices supported by D2DMHRs and the D2DSHRs operating as Cluster Heads (CHs). This info is included in the ProSe discovery message sent through LTE Proximity Services, along with their location and D2D mode. Correct selection of the threshold values, achieves a more efficient and quicker CH selection, providing in this manner improvements in SE and PC.

6.1.4 PROCESS OF WDR CALCULATION AND TRANSMISSION MODE SELECTION USING INITIAL DAIS DESIGN

In this section, we provide the initial version (non-enhanced) of DAIS. So, once a D2D device enters the D2D network for the first time, the DAIS Plan goes through the following steps:

Step 1: The WDR of the path associated with the direct link between the D2D device and the BS is estimated (using WDR_0, Eq. 6.3). This will be compared with the other candidate indirect paths identified in the step below.

$$\text{WDR}_0(\text{D2D}) = B \times \text{SE}(\text{D2D}, \text{BaseStation}) \tag{6.3}$$

where B is the Bandwidth and

SE is the Spectral Efficiency, given in Eq. 7.1

Step 2: Other candidate indirect paths between the D2D device and the BS are identified, their associated WDR is considered, and the best path (i.e., the one with the highest WDR) is selected. Based on the path selected (i.e., direct or indirect path), the Transmission Mode of the D2D device is selected. The details are provided below:

a. Using LTE Proximity Services the entering D2D device scans the network for any neighbouring D2D-Relay devices in order to identify existing D2D communication paths and acquire their WDRs. The broadcast LTE proximity advertisement messages also include additional information, such as the number of D2D devices serviced by the D2D-Relay device and the device that each D2D-Relay connects to next, along the path to the BS/GW (labelled D2D-Relay Next).
b. Using Eq. 6.4 the WDR of the best path is identified.

$$\text{WDR}_{\text{Path with maximum WDR}}(\text{D2D}) = \max_{x=\text{Relay}1,\ldots,\text{Relay}N} (\text{WDR}_{\text{Min Path}}(x)) \tag{6.4}$$

where Relay1...RelayN is the set of D2D-R around a D2D

$$\text{WDR}_{\text{Min Path}}(\text{D2D}) = \min_{y=U_1,\ldots,U_N \in \text{Path}(y)} (\text{WDR}_{\text{Link Data Rate}}(y, y+1))$$

Path(y) returns all the D2D-R in the path towardsBS/Gateway

$$\text{WDR}_{\text{Link Data Rate}}(\text{D2D}, \text{D2D2}) = B \times \text{SE}(\text{D2D}, \text{D2D2})$$

Then, based on received LTE ProSe discovery messages, the state of the nearby D2D devices are classified by the DAIS algorithm (Algorithm 6.1), with the use of Eq. 6.4, into six possible states:

State 1: The D2DSHR (single hop D2D-R) with the maximum WDR within Wi-Fi range (maxD2DSHR).

State 2: The D2DMHR (multihop D2D-R) with the maximum WDR within the range of Wi-Fi Direct and with no connections (maxD2DMHRNoConnections).

State 3: The D2DSHR (single hop D2D-R) with the maximum WDR within the range of LTE Direct and with no connections (maxD2DSHRNoConnectionsToBeD2DMHR).

State 4: The D2DSHR (single hop D2D-R) with the maximum WDR within the range of LTE Direct, with no connections and worst WDR than the entering device (maxD2DSHRToUse UED2DMHR).

State 5: The D2DMHR (multihop D2D-R) with the maximum WDR within the range of LTE Direct and with no connections (maxD2DMHRToUseAsMultiHop).

State 6: If none of the above states is satisfied, the D2D device remains connected to the BS.

c. Transmission Mode Assignment (as shown in Algorithm 6.1): In order for the Transmission Mode Assignment to achieve correct selection of Transmission Mode, it must satisfy the WDR threshold and the BPL threshold. Thus, an entering D2D device, by considering the info included in the LTE ProSe messages, computes its WDR Threshold (note that the BPL is fixed). In the estimation of this threshold, the aggregated number of D2D devices served by the surrounding D2D-Relays,[6] is also considered. The equation used for the computation of the WDR Threshold appears in Section 6.1.3.1.
Once the state and the thresholds are identified, the assignment of the Transmission Mode to the entering D2D device is carried out. In order not to violate the quality checks, the entering D2D device uses the DAIS algorithm to select its Transmission Mode by considering Eqs. 6.5–6.9. Specifically, the execution of DAIS includes some steps that are executed in order, with execution progressively moving to the next step only if the current step is not satisfied. These steps are:
   i. The entering D2D device sets the Transmission Mode to be "D2D Client" and selects to connect to the maxD2DSHR (state 1) if Eq. 6.5 is satisfied.

$$(\mathrm{WDR}_{\mathrm{Threshold}} + 1) \times \mathrm{WDR}_0(\mathrm{D2D}) \leqslant \mathrm{WDR}(\mathrm{maxD2DSHR}) \tag{6.5}$$

   ii. The entering D2D device sets the Transmission Mode to be "D2D Client", selects to connect to the maxD2DMHRNoConnections (state 2) and informs the maxD2DMHRNoConnections device to change its Transmission Mode to D2DSHRfrom D2DMHR if Eq. 6.6 is satisfied.

$$(\mathrm{WDR}_{\mathrm{Threshold}} + 1) \times \mathrm{WDR}_0(\mathrm{D2D}) \leqslant \mathrm{WDR}(\mathrm{maxD2DMHRNoConnections}) \tag{6.6}$$

   iii. The entering D2D device sets the Transmission Mode to be "D2DSHR", selects to connect to the maxD2DSHRNoConnectionsToBeD2DMHR (state 3) and informs the maxD2DSHRNoConnectionsToBeD2DMHR device to change its Transmission Mode to D2DMHR from D2DSHR if Eq. 6.7 is satisfied.

$$\mathrm{WDR}(\mathrm{maxD2DSHRNoConnectionsToBeD2DMHR}) \geqslant (\mathrm{WDR}_{\mathrm{Threshold}} + 1) \times \mathrm{WDR}_0(\mathrm{D2D}) \tag{6.7}$$

iv. The entering D2D device sets the Transmission Mode to be "D2DMHR", selects to connect as a sharing device to the maxD2DSHRToUseUED2DMHR (state 4), and informs the maxD2DSHRToUseUED2DMHR device to connect to the entering D2D device and keep its Transmission Mode to D2DSHR, if Eq. 6.8 is satisfied. In this case, the D2D entering device "breaks" an existing connection.

$$\mathrm{WDR}(\mathrm{maxD2DSHRToUseUED2DMHR}) \leqslant (\mathrm{WDR}_{\mathrm{Threshold}} - 1) \times \mathrm{WDR}_0(\mathrm{D2D}) \quad (6.8)$$

v. The entering D2D device sets the Transmission Mode to be "D2DSHR" and selects to connect to the maxD2DMHRToUse AsMultiHop (state 5), if Eq. 6.9 is satisfied.

$$\mathrm{WDR}(\mathrm{maxD2DMHRToUseAsMultiHop}) \geqslant (\mathrm{WDR}_{\mathrm{Threshold}} + 1) \times \mathrm{WDR}_0(\mathrm{D2D}) \quad (6.9)$$

vi. The entering device sets the Transmission Mode to be "D2DMHR" and selects to connect to the BS (state 6).

d. The entering D2D device sets the selected Transmission Mode.

Step 3: As a final step, the WDR is assigned as the minimum value among the link data rate between the entering D2D device towards the selected D2D-Relay as shown in Eq. 6.10.

$$\mathrm{WDR}_{\mathrm{assigned}}(\mathrm{D2D}) = \min(\mathrm{WDR}_{\mathrm{Link\ DataRate}}(\mathrm{D2D}, z), \mathrm{WDR}(z)) \quad (6.10)$$

where z is the D2D-R node selected

In our approach, the D2D-Relays are using proximity services to broadcast their connection information (i.e. WDR, coordinates).

6.1.5 ENHANCEMENTS OF DAIS

Based on our findings with the initial DAIS plan, it became evident that the WDR and BPL levels are affected by the number of D2D devices, hence influencing the behaviour of DAIS. Thus, DAIS is enhanced with: (i) an additional parameter (number of D2D devices in proximity) to be considered in the transmission mode selection; and (ii) the dynamically settable WDR and BPL thresholds, adapted and fine-tuned for scenarios with a range from 10 to 1000 UEs. Consequently, the major DAIS enhancement provided to a D2D device (newly entering or existing) is the ability to dynamically set and use threshold values for the WDR and the BPL. These values are dependent on the number of other D2D devices in a specific radius around the device, and they are selected so as to provide the best results in terms of SE and PC. The radius depends on the coverage range of the broadcast ProSe message. Note extensive evaluation and selection of best threshold[7] values was carried out. A detailed reporting of this work appears in Ref. [4] for a small number of devices and in Ref. [5] for a large number of devices.

Algorithm 6.1 Transmission Mode Selection and Cluster Formation by Utilizing WDR (DAIS)

```
 1: i: radius of Selecting Device Around UE
 2: WDR: my WDR to BS
 3: T: a set containing all D2D devices information (i.e. Data Rate,Coordinates) from all local network, provided by
    LTE ProSe
 4: procedure TRANSMISSIONMODESELECTIONWITHWDR(T_th, i, DR)
 5:     calculate from T_th maxD2DSHR,maxD2DMHRNoConnections,
 6:     maxD2DSHRNoConnectionsToBeD2DMHR,
 7:     maxD2DSHRToUseUED2DMHR,
 8:     maxD2DMHRToUseAsMultiHop
 9:     if ∃ maxD2DSHR then
10:         Connect UE as D2D Client to maxD2DSHR using Wi-Fi Direct
11:     else if ∃ maxD2DMHRNoConnections then
12:         Request from maxD2DMHRNoConnections UE to be D2DSHR
13:         Connect UE as D2D Client to maxD2DMHRNoConnections using Wi-Fi Direct
14:     else if ∃ maxD2DSHRNoConnectionsToBeD2DMHR then
15:         Request from maxD2DSHRNoConnectionsToBeD2DMHR UE to be D2DMHR
16:         Connect UE as D2DSHR to maxD2DSHRNoConnectionsToBeD2DMHR using LTE Direct
17:     else if ∃ maxD2DSHRToUseUED2DMHR then
18:         Set UE as D2DMHR
19:         Connect maxD2DSHRToUseUED2DMHR as D2DSHR to UE using LTE Direct
20:     else if ∃ maxD2DMHRToUseAsMultiHop then
21:         Set UE as D2DSHR
22:         Connect UE as D2DSHR to maxD2DMHRToUseAsMultiHop using LTE Direct
23:     else
24:         set UE as D2DMHR
25:         Stay connected to BS
26:     end if
27: end procedure
```

6.1.6 EXTENDED DAIS TO HANDLE THE DYNAMIC ENVIRONMENT ASPECTS

To handle dynamic situations, we extend the DAIS plan to achieve transmission mode selection by considering the dynamics of the Mobile Network causing variations in the D2D network topology. These relate to changes in UE speed, UE direction, number of devices in a D2D communication network, etc. Our target is to extend the DAIS plan to achieve better SE and PC, in a dynamic mobile environment, by dynamically re-forming the connections and clusters. Thus, the enhanced DAIS plan, initially introduced in Section 6.1 and enhanced in Section 6.1.5, is extended targeting the creation of stable and efficient clusters and good backhauling links towards the gateway, considering dynamic network conditions through subsequent Time Steps (TS) of execution. To achieve this, the algorithm of enhanced DAIS plan is extended with the Speed Threshold (named "MAXSpeedToFormBackhauling"; see Appendix) in the decision process with a value of 1.5 m/s (pedestrian speed), which allows a device to be a D2D-Relay if its speed is lower than the threshold. The difficulty in the dynamic environment is that in each Time Step of execution the new selected Transmission Mode can affect existing clusters, as well as the formation of new clusters and backhauling links, which could result in disconnected/disjointed clusters. However, these clusters and paths should not be affected, even if the UE moves away from the Cluster Head (CH).

The dynamic DAIS implementation Plan is shown in Algorithm 6.2, using for the BPL Threshold (DeviceBatteryThreshold) a value of 75%, and for the WDR Threshold (PERCDataRate), a value that is dependant on the number of D2D Devices in the network (i.e., ≤200 20% WDR threshold value, >200 35% WDR threshold value; see Section 6.1.5). The number of D2D Devices in the network is made known through LTE ProSe messages that the D2D-Relays share with all other devices, incorporating in the message the number of the clients they serve.

Algorithm 6.2 DAIS Algorithm for Transmission Mode Selection Plan in BDIx Agents

```
 1: i: radius of Selecting Device Around UE
 2: WDR: my WDR to BS
 3: speed: the speed of D2D
 4: DeviceBatteryThreshold : 75%
 5: PERCDataRate: 20% for < 200 D2D Devices or else 35%
 6: T: a set containing D2D-Relay information (i.e., WDR, Coordinates, Number of Devices) from all network, pro-
    vided by using ProSE messages
 7: procedure TRANSMISSIONMODESELECTIONWITHWDR(T_th, i, DR)
 8:     calculate from T_th maxD2DSHR,maxD2DMHRNoConnections,
 9:     maxD2DSHRNoConnectionsToBeD2DMHR,
10:     maxD2DSHRToUseUED2DMHR,
11:     maxD2DMHRToUseAsMultiHop
12:     WeightedDataRateSelectedD2DSHR = Link Weighted Data Rate among WDR and maxD2DSHR
13:     if ∃ maxD2DSHR ∧ WeightedDataRateSelectedD2DSHR≥(1.0+PERCDataRate)*WDR then
14:         Connect UE as D2D Client to maxD2DSHR using Wi-Fi Direct
15:     else if ∃ maxD2DMHRNoConnections then
16:         Request from maxD2DMHRNoConnections UE to be D2DSHR
17:         Connect UE as D2D Client to maxD2DMHRNoConnections using Wi-Fi Direct
18:     else if ∃ maxD2DSHRNoConnectionsToBeD2DMHR∧ speed<MAXSpeedToFormBackhauling ∧
    battery>DeviceBatteryThreshold then
19:         Request from maxD2DSHRNoConnectionsToBeD2DMHR UE to be D2DMHR
20:         Connect UE as D2D Relay to maxD2DSHRNoConnectionsToBeD2DMHR using LTE Direct
21:     else if ∃ maxD2DSHRToUseUED2DMHR ∧ speed<MAXSpeedToFormBackhauling ∧
    battery>DeviceBatteryThreshold then
22:         Set UE as D2DMHR
23:         Connect maxD2DSHRToUseUED2DMHR as D2D Relay to UE using LTE Direct
24:     else if ∃ maxD2DMHRToUseAsMultiHop ∧ speed<MAXSpeedToFormBackhauling ∧
    battery>DeviceBatteryThreshold then
25:         Set UE as D2DSHR
26:         Connect UE as D2D Relay to maxD2DMHRToUseAsMultiHop using LTE Direct
27:     else
28:         set UE as D2DMHR
29:         Stay connected to BS
30:     end if
31: end procedure
```

6.2 DSR FOR TRANSMISSION MODE SELECTION IN STATIC AND DYNAMIC ENVIRONMENT

This section provides the implementation of Sum Rate (SR), a description of the Distributed Sum Rate (DSR), and the calculation of the thresholds adjusted for the DSR and the SR enhancements. Also, this section provides the additional extension implemented at Distributed Sum Rate approach to support a dynamic environment with speed and direction. The extra extension includes the previous enhancements of DSR and additionally the examination of D2D device speed with the speed threshold for allowing the device to be D2D-Relay.

6.2.1 SUM RATE

One of the most common metrics for the evaluation of D2D solutions is the Sum Rate (SR). The SR is the total throughput in a network calculated as the sum of the data rates that are delivered to all UEs and D2D UEs in a network. In the SR approach, when a new device enters the cell, the BS gathers the connections and the Transmission Mode of all the devices and calculates the Transmission Mode of the entering device by executing a brute force investigation for all transmission modes and all connections (according to the thresholds of D2DSHR and D2DMHR) and then selects the Transmission Mode that achieves the maximum SR. Thus, the Sum Rate is a centralized algorithmic maximization approach that selects the transmission mode that the D2D Device will operate by using global network knowledge (i.e., Coordinates, Data Rates, Transmission Modes and Links of all Devices under the BS) and by focusing on maximizing the aggregated data rate of all the links established in the Network. Overall, we consider the Sum Rate approach the best approach because it uses brute force investigation to conclude with the best transmission mode in terms of SE/PC in each D2D Device. In Section 7.1.5 there is a comparison with Sum Rate and the initial DAIS shown in Section 6.1.4, resulting in both to achieve the same SE and PC. However, the DAIS was faster in execution.

6.2.2 DISTRIBUTED SUM RATE

Distributed Sum Rate (DSR) implements a distributed approach where each D2D device selects its Transmission Mode with their target the maximization of the Sum Rate in the network. To achieve this, the approach first calculates all possible cases/combinations that can be achieved related to: (i) the transmission mode that the entering D2D device can select to operate (i.e., D2D Relay, D2D Multi Hop Relay, D2D Client, D2D Direct); and (ii) the link to which D2DSHR/D2DMHR the D2D device will select to connect. Then it selects the case/combination with the highest achieved SR. A brief outline of DSR implementation, as well as a description of the enhancements proposed in this chapter is shown below.

The DSR approach, introduced as Sum Rate at Section 6.2.1, evaluates the maximum SR (in a similar way to Ref. [6]) to achieve the best transmission mode, best link, and best path to the BS or Gateway. In order to allow for a fairer comparison, in this chapter the DSR algorithm is adapted to utilize the terms and parameters (see Appendix A.1), and thresholds (see Section 6.1.3) of DAIS. Furthermore, the adapted DSR is enhanced to use and accommodate the algorithm defined for DAIS, thus providing the ability to an entering D2D device to alter the D2D network structure and either: (i) replace an existing D2D-Relay device and take its role accordingly; or (ii) break an existing sharing connection of a D2D-Relay (with another D2D device), update its Transmission Mode (if needed) and connect with it accordingly. The adapted DSR algorithm is shown in Algorithm 6.3, and is executed whenever a new D2D device enters the D2D communication network.

Algorithm 6.3 Adapted DSR Algorithm for Transmission Mode Selection and Cluster Formation

1: D2D'DSR: The D2D device running the DSR algorithm
2: Radius: Scanning radius of D2D'DSR for locating D2DSHRs,D2DMHRs around it
3: DR: Date Rate of the link between the D2D'DSR and the BS
4: InfoSet: A set including information related to all D2D devices of the D2D Network (i.e. Data Rate, Coordinates, Transmission Mode). This InfoSet is provided by the BS to the D2D'DSR
5: **procedure** TRANSMISSIONMODESELECTIONWITHDSR($\text{InfoSet}_{\text{th}}$, Radius, DR)
6: Call SecurityD2DCommunication(InfoSet, MSISDN, IMEI) ((this algorithm forms part of a separate study))
7: Calculate from $\text{InfoSet}_{\text{th}}$ the following values:
8: maxD2DSHR
9: maxD2DMHRNoConnections
10: maxD2DSHRNoConnectionsToBeD2DMHR
11: maxD2DSHRToUseUED2DMHR
12: maxD2DMHRToUseAsMultiHop
13: **if** ∃ maxD2DSHR **then**
14: Connect D2D'DSR as D2D Client to maxD2DSHR using Wi-Fi Direct
15: **else if** ∃ maxD2DMHRNoConnections **then**
16: Request from maxD2DMHRNoConnections to become a D2DSHR
17: Connect D2D'DSR as D2D Client to maxD2DMHRNoConnections using Wi-Fi Direct
18: **else if** ∃ maxD2DSHRNoConnectionsToBeD2DMHR **then**
19: Request from maxD2DSHRNoConnectionsToBeD2DMHR to become D2DMHR
20: Set D2D'DSR as D2DSHR
21: Connect D2D'DSR to maxD2DSHRNoConnectionsToBeD2DMHR using LTE Direct
22: **else if** ∃ maxD2DSHRToUseUED2DMHR **then**
23: Set D2D'DSR as D2DMHR
24: Set maxD2DSHRToUseUED2DMHR as D2DMHR
25: Connect D2D'DSR to maxD2DSHRToUseUED2DMHR using LTE Direct
26: **else if** ∃ maxD2DMHRToUseAsMultiHop **then**
27: Set D2D'DSR as D2DSHR
28: Connect D2D'DSR to maxD2DMHRToUseAsMultiHop using LTE Direct
29: **else**
30: Set D2D'DSR as D2DMHR
31: D2D'DSR stay connected to BS
32: **end if**
33: **end procedure**

6.2.3 DAIS THRESHOLDS ADJUSTED FOR ENHANCED DSR

The DSR approach uses the BPL and the Link Data Rate (LDR) thresholds. The BPL is used as in the DAIS approach. On the other hand, the LDR threshold uses the same values and a similar approach to the WDR threshold used in DAIS, however it is used differently. Specifically, the LDR threshold is used to compare a value expressed by the ratio of: the Data Rate of the link that will be created, replaced or canceled in the D2D communication network for the entering D2D device, divided by the Data Rate of the existing link of the D2D device with the BS.

The LDR Threshold is used by the DSR algorithm when a new D2D device enters the Network for four purposes:

To perform a quality check of the D2DSHRs in the D2D network (maxD2DSHR in Algorithm 6.3), in order to connect to one of them as a D2D Client. Basically, the new D2D device entering the D2D network, will: (i) acquire from the BS all the D2DSHRs in its proximity which it can connect to as a D2D Client; (ii) Using Eqs.6.11 and 6.12,

$$\mathrm{DR}(\mathrm{D2D}) = \max(\mathrm{DR}(\mathrm{D2D}, \mathrm{DR}_{\text{Best Path}}(\mathrm{D2D})), \mathrm{DR}_{\text{Best Path}}(\mathrm{D2D})) \tag{6.11}$$
$$\text{where } \mathrm{DR}_{\text{Best Path}}(\mathrm{D2D}) = \max_{x=\text{Relay},\dots,\text{Relay}N} (DR(x))$$
$$\text{and Relay} \ldots \text{Relay}N \text{ Set of} \mathrm{D2DSHR}, \mathrm{D2DMHR} \text{ around } \mathrm{D2D}$$
$$\text{where } \mathrm{DR}_{\mathrm{D2D}} = B \times \mathrm{SE}(\mathrm{D2D}, \mathrm{D2D2})$$
$$\text{where } B \text{ is the Bandwidth}$$
$$(\mathrm{LDR}_{\text{Threshold}} + 1) \times \mathrm{DR}_{\mathrm{D2D}} \leqslant \mathrm{DR}_{\text{Best Path}}(\mathrm{D2DSHR}) \tag{6.12}$$

filter the D2DSHRs based on their LDR[8] and the LDR Threshold set; (iii) sort the D2DSHRs in descending order based on the sum of their LDR + Sum Rate; and (iv) select and connect to the D2DSHR with the highest achievable LDR + Sum Rate.

To perform a quality check of the D2DMHRs (maxD2DMHRNoConnections in Algorithm 6.3), in order to connect to one of them either as a D2DMHR or a D2DSHR (this is based on the distance of the D2D device from the D2DMHR Device). The steps followed are the same as above.

To perform a quality check of the entering D2D device, in order to replace the D2DSHR (maxD2DSHRNoConnectionsToBeD2DMHR in Algorithm 6.3) or D2DMHR (maxD2DSHRToUseUED2DMHR in Algorithm 6.3) in D2D network. Basically, the new D2D device entering the D2D network, will: (i) extract from the information sent by the BS, all the D2D-Relay in its proximity which can connect to as D2D-R; (ii) sort the D2DSHRs in descending order based on the sum of their LDR + Sum Rate; (iii) Using Eq.6.13,

$$\mathrm{D2D}_{\text{Share}} \leqslant (\mathrm{LDR}_{\text{Threshold}} - 1) \times \mathrm{DR}_{\mathrm{D2D}} \tag{6.13}$$
$$\text{where } \mathrm{D2D}_{rmShare} \text{ is } \mathrm{DR}_{\text{Best Path}}(\mathrm{D2DSHR})$$
$$\vee \mathrm{DR}_{\text{Best Path}}(\mathrm{D2DMHR})$$

filter the D2D-Relays based on their LDR and the LDR Threshold set; and (iv) select and replace the first D2D-Relay that has the highest achievable LDR + Sum Rate as D2D-Relay according to the algorithm.

To perform a quality check of the D2D-Relay in the D2D network (maxD2 DMHRToUseAsMultiHop in Algorithm 6.3), in order to connect to one of them as a D2DSHR. More specifically, the new D2D device entering the D2D network, will: (i) extract from the information sent by the BS all the D2D-Relays with no connection in its proximity which can connect to as a D2D-Relay;

(ii) Using Eq.6.14,

$$D2D_{Share} \geqslant (LDR_{Threshold} + 1) \times DR_{D2D} \tag{6.14}$$

$$\text{where } D2D_{Share} \text{ is } DR_{Best\ Path}(D2DSHR) \vee DR_{Best\ Path}$$

and is the Link of D2DSHR, D2DMHR to the Entering D2D

filter the D2D-Relays based on their LDR and the LDR Threshold set; (iii) sort the D2D-Relays in descending order based on the sum of their LDR + Sum Rate and select the one with the highest LDR + Sum Rate; and (iv) if the one selected is D2DMHR, then the D2D device will become a D2DSHR and connect to it. Otherwise, if the one selected is D2DSHR then it will change its transmission mode into D2DMHR and the D2D device will become a D2DSHR and connect to it.

6.2.4 ENHANCED DSR ALGORITHM FOR TRANSMISSION MODE SELECTION

For the enhanced DSR, the execution of the DSR algorithm is moved from the BS to the D2D devices and realized in a distributed manner. Additionally, DSR is enhanced[9] with the accommodation of the same thresholds as DAIS for the static environment, and the ability of a D2D device to alter existing links (similar to DAIS functionality). These enhancements achieve, for both approaches, high impact on the selection of the cluster heads and the formation of more efficient clusters, in terms of SE and PC. Algorithm 6.3 provides the steps performed by the DSR approach (extensively enhanced/adjusted from the one proposed in Section 6.2.1) for the transmission mode selection and the formation of the clusters. The terms and parameters used for DAIS, but also utilized and used for DSR, are provided in Appendix A.1. The DSR algorithm is activated when a UE (capable to perform D2D communication) enters the D2D network. The aim is to select the transmission mode that the UE will operate in the D2D network. Depending on the DSR decision, the UE might connect to the D2D network as D2D Client, D2DSHR or D2DMHR (either connected with the BS or, as a bridge, with another D2DMHR or D2DSHR), altering in this way the D2D network structure.

6.2.5 EXTENDED DSR TO HANDLE THE DYNAMIC ENVIRONMENT ASPECTS

This section introduces the Speed Threshold as an extension in the enhanced DSR approach, to make it competitive, distributed and align with DAIS in a dynamic environment. The DSR is adapted and extended from the Sum Rate approach to use and accommodate the algorithm defined for DAIS (shown in Section 6.1.6), and utilize the same terms, parameters, and some of its thresholds (i.e., BPL Threshold, Speed Threshold) as shown in Section 6.2.4. This provided the ability to the Sum Rate approach to operate in a distributed manner and allow an entering D2D device to alter the D2D network structure.[10] The implementation of the extended DSR is shown in the Algorithm 6.4 and is executed whenever a new D2D device enters the D2D communication network.

Similarly to the extended DAIS approach described above, for the Speed Threshold, called "MAXSpeedToFormBackhauling", we consider a pedestrian speed (i.e., 1.5 m/s). The BPL threshold, called "DeviceBatteryThreshold", uses a value of 75% is used. To this end, the extended DSR approach assigns the D2D-Relay Transmission Mode only in devices that achieve the above thresholds. Additionally, we use the Threshold for Sum Rate (set empirically to 35%) called "DataRateThreshold". This threshold is used for a quality check when a Device attempts to connect as a client to a D2D-Relay Device. More specifically, for a D2D candidate device to connect as a D2D client at a D2D Relay, the client's data rate must be at least equal to 1.35*Data Rate to BS, where the Data Rate to BS is the Data rate of the D2D candidate device towards the BS. If the aforementioned threshold is not satisfied, the D2D candidate device will select the next best Transmission Mode (i.e., either as D2DSHR or D2DMHR) that achieves the maximum Sum Rate.

Algorithm 6.4 Sum Rate Algorithm for Transmission Mode Selection in Extended DSR Approach

1: i: radius of Selecting Device Around UE
2: DR: my Data Rate to BS
3: speed: the speed of D2D
4: battery: the battery Level of D2D
5: DeviceBatteryThreshold: 75%
6: DataRateThreshold: 35%
7: T: a set containing D2D-Relay information (i.e. Data Rate, Connections, Coordinates, Number of Devices) from all network, provided by BS through message exchange.
8: **procedure** TRANSMISSIONMODESELECTION(T_{th}, i, DR)
9: calculate from T_{th} existingNetworkSumRate,
10: max SumRateIfSelectD2DMHR to a
11: D2DMHRSelectedD2DMHRorBS,
12: max SumRateIfSelectD2DSHR to a D2DSHRSelectedD2DMHRorBS,
13: max SumRateIfSelectD2DClient to a SelectedD2DSHR
14: DataRateSelectedD2DSHR = Link Data Rate among CanditateD2D and SelectedD2DSHR
15: **if** $\exists$ SumRateIfSelectD2DClient is maximum Sum Rate $\wedge$ DataRateSelectedD2DSHR $\geq$ (1.0+DataRateThreshold)*DR **then**
16: Connect UE as D2D Client to SelectedD2DSHR using Wi-Fi Direct
17: **else if** $\exists$ SumRateIfSelectD2DSHR is maximum Sum Rate $\wedge$ speed $<$ MAXSpeedToFormBackhauling $\wedge$ battery $>$ DeviceBatteryThreshold **then**
18: Connect UE as D2D Relay to D2DSHRSelectedD2DMHRorBS using LTE Direct
19: **else if** $\exists$ SumRateIfSelectD2DMHR is maximum Sum Rate $\wedge$ speed$<$MAXSpeedToFormBackhauling $\wedge$ battery$>$DeviceBatteryThreshold **then**
20: Connect UE as D2D Multi Hop Relay to D2DMHRSelectedD2DMHRorBS using LTE Direct
21: **else**
22: set UE as D2DMHR
23: Stay connected to BS
24: **end if**
25: **end procedure**

6.3 ENHANCED SINGLE HOP RELAY APPROACH USED IN DYNAMIC ENVIRONMENT

In order to be fair in our investigation, we enhanced Single Hop Relay Approach (SHRA), introduced in Ref. [7], in order to support multiple connections at D2D-Relays and allow cluster formation and to consider the same parameters as the other investigated approaches examined in a dynamic environment. The SHRA approach is enhanced in our investigation in the sense that the D2D Relay accepts more than one connection and serves as a regular D2D Relay, rather than an intermediate D2D Device, as the author suggests. As with the previous approaches, the SHRA is modified to use Wi-Fi Direct when selecting D2D Relay with the limitation of distance to clients to 200 m and the limitation of the number of clients to 200. The D2D connection distance among two D2D Devices is the same as it was defined in the investigated section to the value of "30 meters" as in Ref. [7]. Additionally, in this approach, we consider that each D2D device in the network uses LTE ProSe to share its coordinates and transmission mode with all other devices. By considering mobility, these improvements are implemented within the approaches mentioned above, providing enhanced performance in terms of SE and PC and reduced computation time (as shown in Section 7.2.2.3).

6.4 DISTRIBUTED RANDOM AND NON-D2D UE FOR TRANSMISSION MODE SELECTION

The Distributed Random (DR) approach is a simple approach that selects the Transmission Mode of each UE in a random manner. The transmission mode selection is performed in a distributed manner using the global network knowledge (acquired from the BS) and depends on the number of D2D devices in the network. Note that DR acquires only the D2DSHR and D2DMHR near the D2D Candidate Device according to constraints. The non-D2D UE approach describes the current approach used in Mobile Networks. This approach keeps all the UEs connected directly to the BS and a constant predefined transmission power, that is specified for the UEs that are directly connected to the BS, is used.

We consider the Distributed Random to be the worst approach that results in the worst SE. Similarly, this investigation considers the non-D2D UE approach to be the worst method in terms of PC.

6.5 HEURISTIC ALGORITHM FOR ADAPTING THE CLUSTERING RESULTS OF FUZZY ART, DBSCAN, G-MEANS, AND MEC

It is important to highlight here that Fuzzy ART, DBSCAN, G-MEANS and MEC clustering techniques were not designed for application in D2D communication specifically. Thus, to allow a fairer comparative performance evaluation a heuristic algorithm (the Algorithm 6.5) was developed with the aim to adapt their clustering results so as to operate for D2D communication.

Note that Fuzzy ART, DBSCAN, MEC and G-MEANS (Section 2.5.5) are centralized unsupervised learning AI/ML clustering techniques, which we adapted for the purposes of this research (Section 6.5) in order to operate for D2D communication. The aforesaid unsupervised clustering techniques are selected for the comparative performance evaluation as they: (i) can perform Cluster Head selection with the use of a heuristic algorithm, which is directly associated with the Transmission Mode that will used by the D2D devices; (ii) do not require a learning process in order to perform clustering. This is an important aspect for D2D networks as they are mostly dynamic in nature due to the mobility of the D2D devices; (iii) are not demanding in terms of memory or CPU power, thus they do not burden the BS or the D2D device; (iv) provide good clustering results in short time; and (v) are well used and well known for finding clusters in similar problems (e.g., clustering of system alerts, clustering of security attacks). Moreover, to gain further insight into their performance, we introduced three metrics in terms of SE (Spectral Efficiency) and Power Consumption (PC): D2D effectiveness, D2D Stability, and D2D productivity (Section 7.1.3).

An outline of the steps followed is: (i) An unsupervised learning clustering algorithm (i.e., Fuzzy ART, DBSCAN, G-MEANS or MEC) is first executed and groups all UEs within the coverage area of the BS into clusters based on location; (ii) The clusters formed (we refer to these as CSet in Algorithm 6.5) are provided as input in our Heuristic algorithm; (iii) For each cluster formed, the heuristic algorithm identifies the UE that will become D2DSHR and the CH of the Cluster (i.e., the UE that has the highest data rate with the BS); and (iv) For each cluster formed, the heuristic algorithm identifies the UEs that will connect to the selected CH and sets them as D2D Clients (i.e., UEs with Euclidean Distance between them and the CH less than the Radius of the CH; for Wi-Fi Direct, this radius is equal to 200 m).

Algorithm 6.5 Heuristic Algorithm Used to Select the Cluster Heads and Form the Clusters

1: Radius: Radius of the Cluster Head
2: CSet: A set containing UEs organized into clusters
3: **procedure** CLUSTERHEADDETECTIONANDDEVICEASSIGMENT($CSet_{th}$, Radius)
4: $CSet_{u_{Radius}} \leftarrow$ list of Clusters from $CSet_{th}$
5: **for** each cluster c in $CSet_{u_{Radius}}$ **do**
6: Node $c_{Radius} \leftarrow$ the UE with maximum Data Rate in cluster c
7: Nodes $c_{Radius} \leftarrow$ list of UEs of cluster c
8: **for** each node n in Nodes c_{Radius} **do**
9: $d(n, \text{Node } c_{Radius}) = \sqrt{\sum_{j=1}^{2} (n_j - \text{Node } c_{Radius\,j})^2}$
10: **if** $d(n, \text{Node } c_{Radius}) \leq$ Radius **then**
11: $n \leftarrow$ Cluster Head Node c_{Radius}
12: **end if**
13: **end for**
14: **end for**
15: **end procedure**

Note that the MEC approach needs to be initialized with results extracted by another clustering approach (as shown in Section 2.5.5). For this case we used K-Means. Moreover, in order to apply the Fuzzy ART, DBSCAN, G-MEANS and MEC approaches to the needs of D2D Communication, we set the constraints/settings as below:

For all approaches, we set the maximum distance to form a cluster at a radius of 200 m (Wi-Fi Direct).
For Fuzzy ART we do not limit the maximum number of clusters allowed (maxClusterCount = −1).
For DBSCAN we set the minimum points (minPts) of the cluster to 2.
For G-MEANS and MEC we set the number (kmax) of clusters (k) to 1000.

It is worth indicating here, that except from the aforesaid constraints/settings set for the AI/ML approaches, all other default settings and constraints provided by the "SMILE" framework are the same [8].

6.6 COMPARISON OF DAIS AND DSR WITH THE APPROACHES SHOWN IN THE RELATED WORK ON TRANSMISSION MODE SELECTION IN D2D COMMUNICATION AT A STATIC ENVIRONMENT

In this section we compare the approaches [9–22] that are related to transmission mode selection in D2D communication shown in the Section 2.5.3 with the DAIS and DSR.

It is worth mentioning that all the investigated approaches have as execution outcome two categories of UEs. In the first category, the selected UEs become part of the D2D network. In the second category, the UEs do not consider entering the D2D network and thus may lose all the advantages of the D2D network (e.g., better SE, less PC) by staying connected to BS as regular UEs. On the other hand, our proposed DAIS (shown in Section 6.1.5) and DSR (shown in Section 6.2.4) approaches consider all the UEs as candidates to become a D2D device. By doing this, compared to the other investigated approaches, DAIS and DSR achieve much better network performance in terms of SE and PC. All approaches feature tradeoffs in terms of signalling overhead and control delay in responding to changes, as discussed below.

Enhanced DSR (shown in Section 6.2.4) performs better than Enhanced DAIS(shown in Section 6.1.5) in terms of SE and PC, but as a distributed approach based on global knowledge, necessitates additional signalling overhead and results in delayed control decisions. On the other hand, DAIS, which relies only on local knowledge, operates with reduced signalling overhead and much faster control decision updates (less than 100 ms). Furthermore, to the best of our knowledge, there is not any other approach in the open literature that tackles the problem of having a D2D device utilizing all transmission modes (D2DSHR, D2D multi-hop and D2D cluster) in a distributed manner, as DAIS and DSR approaches do. Additionally, DAIS and DSR, by introducing and utilizing the WDR and SR metrics, respectively, achieve D2D transmission mode selection in a more efficient manner (see Section 7.1.6).

REFERENCES

1. C. Yang, X. Xu, J. Han, and X. Tao, "GA based user matching with optimal power allocation in D2D underlaying network," *IEEE Vehicular Technology Conference*, pp. 1–5, Las Vegas, Jan. 2014.
2. C. Xu, L. Song, Z. Han, Q. Zhao, X. Wang, X. Cheng, and B. Jiao, "Efficiency resource allocation for device-to-device underlay communication systems: A reverse iterative combinatorial auction based approach," *IEEE Journal on Selected Areas in Communications*, vol. 31, no. 9, pp. 348–358, 2013.
3. R. Wang, J. Zhang, S. H. Song, and K. B. Letaief, "QoS-aware channel assignment for weighted sum-rate maximization in D2D communications," *2015 IEEE Global Communications Conference, GLOBECOM 2015*, pp. 1–6, San Diego, CA, Dec. 2015.
4. I. Ioannou, C. Christophorou, V. Vassiliou, and A. Pitsillides, "5G D2D transmission mode selection performance & cluster limits evaluation of distributed AI and ML techniques," in *2021 IEEE International Conference on Communication, Networks and Satellite (COMNETSAT)*, pp. 70–80, Online Conference, 2021.
5. I. Ioannou, C. Christophorou, V. Vassiliou, and A. Pitsillides, "Performance evaluation of transmission mode selection in D2D communication," *NTMS 2021 Conference*, Paris, Jan. 2021.
6. Z. Zhou, M. Dong, K. Ota, J. Wu, and T. Sato, "Energy efficiency and spectral efficiency tradeoff in device-to-device (D2D) communications," *IEEE Wireless Communications Letters*, vol. 3, no. 5, pp. 485–488, 2014.
7. U. N. Kar and D. K. Sanyal, "Experimental analysis of device-to-device communication," *2019 12th International Conference on Contemporary Computing, IC3 2019*, pp. 1–6, Noida, Aug. 2019.
8. "Smile - Statistical Machine Intelligence and Learning Engine." [Online]. Available at: https://haifengl.github.io/, Accessed on: 2020-03-28.
9. K. Doppler, C. H. Yu, C. B. Ribeiro, and P. Jänis, "Mode selection for device-to-device communication underlaying an LTE-advanced network," *IEEE Wireless Communications and Networking Conference, WCNC*, Sydney, Apr. 2010.
10. M. Jung, K. Hwang, and S. Choi, "Joint mode selection and power allocation scheme for power-efficient device-to-device (D2D) communication," *IEEE Vehicular Technology Conference*, pp. 1–5, Yokohama, May 2012.
11. H. Pang, P. Wang, X. Wang, F. Liu, and N. N. Van, "Joint mode selection and resource allocation using evolutionary algorithm for device-to-device communication underlaying cellular networks," *Journal of Communications*, vol. 8, no. 11, pp. 751–757, 2013.
12. M. H. Han, B. G. Kim, and J. W. Lee, "Subchannel and transmission mode scheduling for D2D communication in OFDMA networks," *IEEE Vehicular Technology Conference*, pp. 1–5, Quebec City, Sep. 2012.
13. S. Xiang, T. Peng, Z. Liu, and W. Wang, "A distance-dependent mode selection algorithm in heterogeneous D2D and IMT-advanced network," *2012 IEEE Globecom Workshops, GC Wkshps 2012*, pp. 416–420, Anaheim, CA, Dec. 2012.
14. Y. Liu, "Optimal mode selection in D2D-enabled multibase station systems," *IEEE Communications Letters*, vol. 20, no. 3, pp. 470–473, 2016.
15. C. Xu, J. Feng, B. Huang, Z. Zhou, S. Mumtaz, and J. Rodriguez, "Joint relay selection and resource allocation for energy-efficient D2D cooperative communications using matching theory," *Applied Sciences (Switzerland)*, vol. 7, no. 5, pp. 1–24, 2017.

16. X. Ma, R. Yin, G. Yu, and Z. Zhang, "A distributed relay selection method for relay assisted device-to-device communication system," *IEEE International Symposium on Personal, Indoor and Mobile Radio Communications, PIMRC*, pp. 1020–1024, Sydney, Sep. 2012.
17. B. Ma, H. Shah-Mansouri, and V. W. Wong, "A matching approach for power efficient relay selection in full duplex D2D networks," *2016 IEEE International Conference on Communications, ICC 2016*, pp. 1–6, Kuala Lumpur, May 2016.
18. M. Zhao, X. Gu, D. Wu, and L. Ren, "A two-stages relay selection and resource allocation joint method for D2D communication system," *IEEE Wireless Communications and Networking Conference, WCNC*, pp. 1–6, Doha, Sep. 2016.
19. H. Feng, H. Wang, X. Chu, and X. Xu, "On the tradeoff between optimal relay selection and protocol design in hybrid D2D networks," *2015 IEEE International Conference on Communication Workshop, ICCW 2015*, pp. 705–711, London, Jun. 2015.
20. L. Wang, T. Peng, Y. Yang, and W. Wang, "Interference constrained D2D communication with relay underlaying cellular networks," *IEEE Vehicular Technology Conference*, pp. 1–5, Las Vegas, NV, Sep. 2013.
21. T. Kim and M. Dong, "An iterative Hungarian method to joint relay selection and resource allocation for D2D communications," *IEEE Wireless Communications Letters*, vol. 3, no. 6, pp. 625–628, 2014.
22. S. M. Kazmi, N. H. Tran, W. Saad, Z. Han, T. M. Ho, T. Z. Oo, and C. S. Hong, "Mode selection and resource allocation in device-to-device communications: A matching game approach," *IEEE Transactions on Mobile Computing*, vol. 16, no. 11, pp. 3126–3141, 2017.

Notes

[1]The D2D client can select Wi-Fi Direct, D2DSHR can share Wi-Fi Direct and connect towards BS with the use of LTE Direct using the assigned frequency from the BS, D2DMHR can share over LTE Direct and connects to the next hop towards the BS using the assigned frequency from the BS.
[2]SE is associated with the sum rate (SR) that can be achieved in the network and the available bandwidth. More specifically, it is the aggregated data rate of all the links established in the Network divided by the available bandwidth of the network.
[3]PC is the aggregated total power used by all the links established in the Network.
[4]The WDR metric represents the minimum data rate of the weakest link in a D2D communication path (directly connected to BS or via D2D-Relay nodes).
[5]In our case, we are not using the overlay of spectrum utilisation. Thus, the Plan "Calculate Best Frequency Band with the least Interference" will return the existing frequency assigned by the BS.
[6]The number of D2D devices served by each D2D-Relay device is included in the LTE ProSe messages.
[7]These two thresholds were initially introduced in DAIS at Section 6.1.3, referred to as PERCDataRate and DeviceBatteryThreshold. A brief explanation of these parameters and their thresholds is discussed in Sections 6.1.3 and 6.1.2.
[8]Date Rate of the new link to be created between the D2D-Relay and the D2D device entering the D2D network.
[9]The enhancements of DSR allow us to further investigate whether DSR, in its distributed form and with extra abilities, has any significant advantages over DAIS.
[10]The entering D2D device can alter the D2D network structure and either: (i) replace an existing D2D-Relay device and take its role accordingly; or (ii) break an existing sharing connection of a D2D-Relay (with another D2D device) update its Transmission Mode (if needed) and connect with it accordingly.

7 DAI Framework Evaluation and Realization for the D2D Mode Selection Challenge

This chapter evaluates the performance of the D2D mode selection challenges in a static and dynamic environment. In the static environment, all nodes in the D2D communication network are in a stationary position, and mode selection is executed by each approach per UE incrementally (i.e., DAIS, DR, DSR, non-D2D UE, Fuzzy ART, DBSCAN, G-MEANS, and MEC). In the dynamic environment, all nodes have speed and direction, resulting in changes at the D2D network topology through subsequent time steps of execution. Thus, mode selection is executed by each approach per UE incrementally and per time step (i.e., DAIS, DR, Sum Rate, non-D2D UE) after the change of each UE position due to the dynamicity of the environment. The different static and dynamic environments selected in the evaluation aim to highlight the DAI framework performance in different situations and its ability to handle this dynamically.

7.1 PERFORMANCE EVALUATION IN A STATIC ENVIRONMENT

This section describes: (i) the evaluation scenarios; (ii) the assumptions and constraints used in the evaluation scenarios; (iii) the introduced evaluation metrics; as well as the commonly adopted metrics of QoS/QoE; and (iv) the simulation environment and its simulation parameters. Additionally, it examines the initial instance (non-enhanced) of the DAIS (shown in Section 6.1.4) with the Sum Rate (shown in Section 6.2.1), DR, non-D2D-UE investigated approaches shown in Section 6.4 as an initial investigation of our book to show how enhancements change the performance of the approaches. Finally, it examines, evaluates, and compares the performance of DAIS and DSR with the unsupervised learning clustering techniques (i.e., Fuzzy-ART, DBSCAN, G-MEANS, and MEC) shown in Section 6.5, DR and non-D2D-UE investigated approaches shown in Section 6.4.

Thus, a comparative performance evaluation of the enhanced DAIS and DSR with several ML unsupervised learning clustering approaches is provided. The performance evaluation aims to investigate the efficiency of DAIS and DSR (in terms of Spectral Efficiency and Power Consumption) compared with other related approaches and identify factors that may affect them, such as link Transmission Power (TP), number of devices in the network, and QoS and QoE considerations. Due to

DOI: 10.1201/9781003469209-7

a lack of other DAI-based D2D transmission mode selection techniques, we adopt several well-known related clustering approaches that can also be exploited for transmission mode selection so as to (indirectly) compare with. Thus, using simulation we compared the performance of the enhanced DAIS and DSR with approaches: (i) Centralized with global view (i.e., Fuzzy ART [1–4], DBSCAN [5–8], G-MEANS [9–11] and MEC [12–14]); and (ii) Decentralized with global view (i.e., Distributed Random (DR) as in Section 6.4).

The performance evaluation considers KPIs provided in Table 2.1, adapted from Ref. [15]. The indicators that it focuses on are: (i) SE; (ii) PC; (iii) execution time; (iv) number of supported UEs by the approach; and (v) configuration time.

7.1.1 ASSUMPTIONS AND CONSTRAINTS

In the performance evaluation, we consider the following assumptions regarding the simulation model:

- A Base Station (BS) with N static, or slow-moving, D2D devices (UEs), where N ranges from 10 to 1000 UEs.
- A connection scenario with a single-antenna and a point-to-point communication.
- A free space path loss model (for calculating average received power). A fading channel model (e.g., Rayleigh, Rician, Nakagami-m) investigation is left for future work.
- A basic noise model, the Additive White Gaussian Noise (AWGN), for calculating the signal-to-noise ratio (SNR).
- Interference is handled by the LTE and Wi-Fi Direct protocols.[1]
- The D2DMHR/D2DSHR transmission modes use a multiple access channel with encoder that can cancel the interference of a UE (as shown in Refs. [20, 21]) after the first transmission in the sharing medium, in any frequency mode (i.e., inband, outband), with the use of Channel State Information(CSI).
- In the D2D multi hop transmission mode, the collaborative D2D devices have enough capacity to achieve the multi hop relay communication, based on the QoS requirements.
- All BDIx agents accept what other agent proposes without considering their Desires/Intentions.

Finally, in our simulation model, we acknowledge that in each D2DMHR node of the back-hauling path we have a penalty for capacity reduction (e.g. in half due to downlink channel). To resolve this issue, a number of technologies can be utilized (i.e., use full-duplex Relays as shown in Refs. [17,22,23], D2D device Wi-Fi and mobile interfaces, hybrid half-duplex/full-duplex scheme as shown in Ref. [24]). Here we assume that one of the aforementioned technologies is enabled for D2DMHR mode.

7.1.2 SIMULATION ENVIRONMENT

We investigate a network with the number of UE devices ranging from 10 to 1000. The devices are placed in a cell range of 1000 m radius from the BS using a Poisson Point Process (PPP) distribution model, with the BS located at the center of the cell. The battery power level of the D2D devices is computed by using a probability estimation function following Gaussian distribution of mean 0.70 and standard deviation 0.30. In our simulation environment we keep the same comparison measurements of performance in all running instances; these are the Total Spectral Efficiency (SE) and Total Power Consumption (PC), achieved by each approach.

It is worth noting here that in our simulation environment each running instance has been simulated ten (10) times using a different PPP distribution model. Thus, the SE and PC values considered for each running instance, which are also provided in the performance evaluation results, correspond to the mean SE and PC values calculated from ten running simulations.

In addition, for the DAIS approach the same simulation constraints, simulation parameters (shown in Table 7.1), formulas for D2D device battery power level estimation and WDR are used, as discussed in Section 6.1.5. Also, the same constraints, simulation parameters and formulas have been utilized by the DSR approach, to allow a fairer comparison (e.g., using similar thresholds for the LSR and BPL). Additionally, both DSR (see Algorithm 6.3 and DAIS (see Algorithm 6.1) implementation algorithms: (i) consider the number of D2D devices in proximity; and (ii) use a different WDR/LSR threshold for small (20%) and large (35%) number of D2D devices (as shown in Refs. [25, 26]) and the same BPL Threshold (75%) for all cases. Also, for the Channel State Information (CSI) we adopt the Statistical CSI. Furthermore, in this investigation, we consider a static scenario, and the time is not involved in any examination.

The simulation environment is implemented using the Java with JADE Framework (it is integrated with FIPA ACL and extended with BDI4JADE library), the LTE/5G Toolbox of MATLAB (2020a) and SMILE (used in AI/ML implementation) libraries. The specs of the machine used for the simulations are as follows: (i) an Intel(R) Core(TM) i7-8750H CPU @ 2.20 GHz; (ii) 24 GB DDR4; (iii) 1TB SSD hard disk; and (iv) NVIDIA GeForce GTX 1050 Ti graphics card with 4GB DDRS5 memory.

7.1.3 EVALUATION METRICS

The performance evaluation considers the KPIs provided in Table 2.1, adapted from Ref. [15], focusing on: (i) SE; (ii) PC; (iii) execution time; (iv) number of supported UEs by the approach; and (v) configuration time. An in-depth evaluation of the investigated approaches will be carried out in terms of Spectral Efficiency (SE) and Power Consumption (PC), while respecting quality criteria. In addition, we also define and consider three new metrics. These metrics are the D2D Effectiveness, the D2D Stability, and the D2D Productivity. Also, the fairness metrics utilized in this investigation are described.

Table 7.1
Simulation Parameters

Simulation Parameters	Value
D2D power	130 mW or otherwise defined [27–29]
UE power	260 mW [27–29]
Wi-Fi direct radius	200 m [30]
LTE direct radius	600 m [31]
BS range	1000 m [27–29]
Path loss exponent (Urban Area)	3.5
BS antenna gain	40 dB [27–29]
UE/D2D antenna gain	2 dB [27–29]
PERCDataRate	20% ($\leq$200) and 35% ($>$200) [25,26]
Device battery threshold	75% [25,26]
No	0.0001 mW
D (max no of D2D clients)	200 Users per Cluster
N (no of UEs)	10–1000
Shadowing	Log-normal
Mobility	Static scenario

7.1.3.1 Spectral Efficiency and Power Consumption

Table 7.2
Parameters Description

Parameter	Parameters Description
C	Capacity (in bits per second bps)
B	Bandwidth (in Hertz Hz)
Si	Signal power (in milli Watts mW)
No	Noise power (in decibel dB or in milli Watts mW)
C_{AWGN}	Capacity with the use of the additive white Gaussian Noise (AWGN) noise model
W	Data bandwidth (in bits per second bps)
SNR	Received Signal-to-Noise Ratio (SNR)
N_0	Noise (in Watts per Herz W/Hz)
$\bar{P}$	Average received power (in mW) (calculated using a free space path loss model)
TP	Transmission Power used by the Device (in mW)

Considering above assumptions and Table 7.2, the SE is derived from the Shannon–Hartley theorem (Eq. 7.1) in (bits/s/Hz).

$$\mathrm{SE} = \frac{C}{B} = \log_2\left(1 + \frac{\mathrm{Si}}{\mathrm{No}}\right) \tag{7.1}$$

Given the Additive White Gaussian Noise (AWGN) as a basic noise model, considering a power- and bandwidth-limited scheme, and a free space path loss model, we calculate the SE from the channel capacity in Eq. 7.2.

$$\begin{aligned} \mathrm{SE} &= \frac{C_{\mathrm{AWGN}}}{W} = \log_2(1+\mathrm{SNR}) \\ \text{where } \mathrm{SNR} &= \frac{\bar{P}}{N_0 W} \end{aligned} \tag{7.2}$$

The PC in mW is given in Eq. 7.3.

$$\mathrm{PC} = \mathrm{TP} - \bar{P} \tag{7.3}$$

The Total SE and Total PC are given below:

$$\text{Total SE} = \sum_{i=1}^{N} \mathrm{SE} \tag{7.4}$$

$$\text{Total PC} = \sum_{j=1}^{N} \mathrm{PC} \tag{7.5}$$

7.1.3.2 D2D Effectiveness, Stability, and Productivity Metrics

To gain further insight into the comparative performance evaluation of the investigated approaches, in terms of SE (Spectral Efficiency) and Power Consumption (PC), we introduced three metrics. These metrics are described below:

a) D2D Effectiveness (%): This metric is used to designate how close to the optimal/best result in terms of SE and PC an approach is, compared to all other investigated approaches. To calculate this metric, first the D2D Ineffectiveness value is computed. Then the D2D Effectiveness value is computed as 1 minus the D2D Ineffectiveness value (as shown in Eq. 7.6 for SE and Eq. 7.7 for PC). It is worth noting that D2D Effectiveness is separated in D2D Effectiveness of SE and D2D Effectiveness of PC. We refer to an approach as D2D SE (PC) Effective if its D2D Effectiveness for SE (PC) is greater than 80% (set empirically). An approach is referred to as D2D Effective if it is both D2D SE and PC Effective.

$$\mathrm{EFF_{SE}}(\mathrm{app}) = 1 - \mathrm{INEFF_{SE}}(\mathrm{app}) \tag{7.6}$$

$$\mathrm{EFF_{PC}}(\mathrm{app}) = 1 - \mathrm{INEFF_{PC}}(\mathrm{app}) \tag{7.7}$$

where

$$\mathrm{EFF_{SE}}(\mathrm{app}) = \bar{S}_{\mathrm{SE}}(\mathrm{app}) = \frac{1}{\mathbf{card}(\mathrm{UEs})} \sum_{\mathrm{UEs}=10}^{1000} \left(1 - S\max_{SE}(\mathrm{UEs}, \mathrm{app})\right)$$

$$\mathrm{EFF_{PC}}(\mathrm{app}) = \bar{P}_{\mathrm{PC}}(\mathrm{app}) = \frac{1}{\mathbf{card}(\mathrm{UEs})} \sum_{\mathrm{UEs}=10}^{1000} \left(1 - P\min_{\mathrm{PC}}(\mathrm{UEs}, \mathrm{app})\right)$$

Considering SE, during each running instance (i.e.number of UEs ranging from 10 to 1000) the D2D Ineffectiveness value of each approach is calculated (in %) as the mean of the SE (see Eq. 7.8), where the difference between the best SE value (i.e., maximum) achieved by all approaches (referred to as the Best SE value) and the SE achieved by the currently investigated approach is divided by the best SE value (as shown in Eq. 7.9), is fed into Eq. 7.8.

$$\mathrm{INEFF}_{\mathrm{SE}}(\mathrm{app}) = \bar{\mathrm{INS}}_{\mathrm{SE}}(\mathrm{app}) = \frac{1}{\mathbf{card}(\mathrm{UEs})} \sum_{\mathrm{UEs}=10}^{1000} S\underset{\mathrm{SE}}{\max}(\mathrm{UEs}, \mathrm{app}) \quad (7.8)$$

$$S\underset{\mathrm{SE}}{\max}(\mathrm{UEs}, \mathrm{app}) = \frac{\max \mathrm{SE}f_{\mathrm{ue}}(\mathrm{UEs}) - f_{\mathrm{se}}(\mathrm{UEs}, \mathrm{TP}_{\max}, \mathrm{app})}{\max \mathrm{SE}f_{\mathrm{ue}}(\mathrm{UEs})} \times 100 \quad (7.9)$$

where
$\max \mathrm{SE}f_{\mathrm{ue}}(\mathrm{UEs}) = \max_{\mathrm{app}=\mathrm{UE},\ldots,\mathrm{MEC}}(f_{\mathrm{se}}(\mathrm{UEs}, \mathrm{TP}_{\max}, \mathrm{app}))$
$\mathrm{app} \in \mathrm{DAIS}, \mathrm{DR}, \mathrm{DSR}, \mathrm{FuzzyART}, \mathrm{DBSCAN}, \mathrm{MEC}, \mathrm{G-MEANS}$
$f_{\mathrm{se}}(\mathrm{UEs}, \mathrm{tra}_{\mathrm{power}}, \mathrm{app}) = \mathrm{SE}_{\mathrm{app}}(\mathrm{UEs}, \mathrm{tra}_{\mathrm{power}})$
$\mathrm{TP}_{\max}$ is the maximum Transmission Power (160mW)
$\mathrm{tra}_{\mathrm{power}} \in \{60, 70, \ldots, 160\}$
$\mathrm{UEs} \in \{10, 20, \ldots, 50, 100, \ldots, 500, 1000\}$
$\mathrm{SE}_{\mathrm{app}}$ Spectral Efficiency of running instance

Similarly for PC, during each running instance (i.e.number of UEs ranging from 10 to 1000) the D2D Ineffectiveness value of each approach is calculated (in %) as the mean of the PC values (see Eq. 7.10), where the difference between the PC value achieved by the investigated approach and the best (i.e., minimum) PC value achieved by all approaches (referred to as Best PC value) is divided by the Best PC value (as shown in the Eq. 7.11), is fed into Eq. 7.10.

$$\mathrm{INEFF}_{\mathrm{PC}} = \bar{\mathrm{INP}}_{\mathrm{PC}}(\mathrm{app}) = \frac{1}{\mathbf{card}(\mathrm{UEs})} \sum_{\mathrm{UEs}=10}^{1000} P\underset{\mathrm{PC}}{\min}(\mathrm{UEs}, \mathrm{app}) \quad (7.10)$$

$$[12pt] P\underset{\mathrm{PC}}{\min}(\mathrm{UEs}, \mathrm{app}) = \frac{f_{\mathrm{pc}}(\mathrm{UEs}, \mathrm{TP}_{\max}, \mathrm{app}) - \min \mathrm{PC}F_{\mathrm{ue}}(\mathrm{UEs})}{f_{\mathrm{pc}}(\mathrm{UEs}, \mathrm{TP}_{\max}, \mathrm{app})} \times 100 \quad (7.11)$$

$$[12pt] \text{where } \min \mathrm{PC}F_{\mathrm{ue}}(\mathrm{UEs}) = \min_{\mathrm{app}=\mathrm{UE},\ldots,\mathrm{MEC}}(f_{\mathrm{pc}}(\mathrm{UEs}, \mathrm{TP}_{\max}, \mathrm{app}))$$

b) D2D Stability: This metric is used to designate the stability of the approach (i.e., how close to the D2D Effectiveness the results are) in terms of SE and PC. For the estimation of this metric, the Standard Deviation[2] of the D2D Effectiveness of the approach, is calculated. The details of how this metric is estimated are given below. It is worth noting that D2D Stability is separated into D2D Stability of SE and D2D Stability of PC. We refer to an approach as D2D SE (PC) Stable if its D2D Stability for SE (PC) is less than 5% (set empirically). An approach is referred to as D2D Stable if it is both D2D SE and PC Stable (as shown in Eq. 7.12 for SE and 7.13 for PC).

$$(\sigma(\hat{app}))^2 = \frac{1}{\mathbf{card}(\text{UEs})-1}\sum_{\text{UE}=10}^{1000}((1-S\max_{\text{SE}}(\text{UE},\text{app}))-\bar{S}_{\text{SE}}(\text{app}))^2 \quad (7.12)$$

$$(\sigma(\hat{app}))^2 = \frac{1}{\mathbf{card}(\text{UEs})-1}\sum_{\text{UE}=10}^{1000}((1-P\min_{\text{PC}}(\text{UE},\text{app}))-\bar{P}_{\text{PC}}(\text{app}))^2 \quad (7.13)$$

c) D2D Productivity: This metric is used to identify the gains or losses of an approach. It is computed by comparing the results (in terms of SE and PC) extracted from the current running instance of the approach with the results extracted from its previous running instance. Again, it is worth noting that D2D Productivity is separated into D2D Productivity of SE and D2D Productivity of PC. We refer to an approach as D2D SE (PC) Productive if its D2D Productivity for SE value (PC value) is greater than 80% (empirically set). An approach is referred to as D2D Productive if it is both D2D SE and PC Productive (as shown in Eq. 7.14 for SE and Eq. 7.18 for PC).
More specifically, in each running instance (i.e.number of UEs ranging from 10 to 1000) the following values related to SE Productivity are calculated:

$$\text{D2D}_{\text{SE}_{\text{Productivity}}} = \frac{\sum_{j=1}^{n}\text{SE}_{\text{gains}}(j)}{n} \times 100 \quad (7.14)$$

$$\text{SE}_{\text{gains}}(j) = \begin{cases} 1\|\text{gains} & \text{if } \text{SE}(\text{UEs},\text{Next UEs},\text{app}) \geq 0 \\ -1\|\text{losses} & \text{if } \text{SE}(\text{UEs},\text{Next UEs},\text{app}) < 0 \end{cases} \quad (7.15)$$

$$\text{SE}(\text{UEs},\text{Next UEs},\text{app}) = \left(\frac{\left(\frac{f_{se}(\text{Next UEs},\text{TP}_{\max},\text{app})}{\text{Next UEs}}\right) - \left(\frac{f_{se}(\text{UEs},\text{TP}_{\max},\text{app})}{\text{UEs}}\right)}{\left(\frac{f_{se}(\text{Next UEs},\text{TP}_{\max},\text{app})}{\text{Next UEs}}\right)}\right) \quad (7.16)$$

$$(7.17)$$

where
Next UEs $\in \{20,\ldots,50,100,\ldots,500,1000\}$
$j \in \{1,\ldots,n\}$ and n is the number of running instances
The SE value (Eq. 7.16) achieved in each running instance by the approach is obtained by estimating the difference between the Average˙SE[3] of the current running instance and the Average˙SE computed in the previous running instance divided by the Average˙SE of the current running instance. In case the computed SE value is positive/negative (Eq. 7.15) the Gains counter is incremented/decremented. Then the SE D2D Productivity is computed by dividing the value stored in the Gains Counter by the total count of running instances (Eq. 7.14).

Following similar arguments, the PC D2D Productivity is computed (as shown in Eq. 7.18):

$$\mathrm{D2D}_{\mathrm{PC_{Productivity}}} = \frac{\sum_{j=1}^{n} \mathrm{PC_{gains}}(j)}{n} \times 100 \tag{7.18}$$

where

$$\mathrm{PC_{gains}}(j) = \begin{cases} 1\|\mathrm{gains} & \text{if } \mathrm{PC(UEs, Next\ UEs, app)} \geq 0 \\ -1\|\mathrm{losses} & \text{if } \mathrm{PC(UEs, Next\ UEs, app)} < 0 \end{cases}$$

$$\mathrm{PC(UEs, Next\ UEs, app)} = -\frac{\left(\frac{f_{pc}(\mathrm{Next\ UEs, TP_{max}, app})}{\mathrm{Next}}\mathrm{UEs}\right) - \left(\frac{f_{pc}(\mathrm{UEs, TP_{max}, app})}{\mathrm{UEs}}\right)}{\left(\frac{f_{pc}(\mathrm{NextUEs, TP_{max}, app})}{\mathrm{Next}}\mathrm{UEs}\right)} \tag{7.19}$$

7.1.3.3 QoE and QoS Fairness Metrics

The investigation performed in this section utilizes two fairness metrics. These are the QoS and the QoE fairness metrics and are used in the performance evaluation in order to quantify and compare the QoE and the QoS fairness provided by each approach. The aforesaid metrics are described below:

(i) The QoS fairness metric can be measured by using the Raj Jain's fairness index (JFI [4]) [32–37]. The equation is provided below.

$$\mathscr{J}(x_1, x_2, \ldots, x_n) = \frac{(\sum_{i=1}^{n} x_i)^2}{n \cdot \sum_{i=1}^{n} x_i^2} = \frac{\overline{\mathbf{x}}^2}{\overline{\mathbf{x}^2}} = \frac{1}{1 + \widehat{c_v}^2}$$

In the aforesaid equation, n is the number of users in the system at a particular instance of time, x_i is the throughput (or any other variable of interest e.g. SE or Data Rate) for the ith connection, and $\widehat{c_v}$ is the sample coefficient of variation (standard deviation/mean). Absolute fairness (i.e., all users receive the same allocation of the shared resources) is achieved when JFI = 1 and absolute unfairness is achieved when JFI = $\frac{1}{n}$. The main reason for selecting JFI as a QoS fairness metric is that JFI is not significantly sensitive to a typical network flow patterns, like D2D communication networks. Also underutilized channels can be identified.

(ii) The QoE fairness metric quantifies fairness among users by considering the Quality of Experience (QoE) as perceived by the end user at the UE device. QoE fairness is considered when the network management aim is to keep the users satisfied in a fair manner. A typical way to measure QoE is by using interval scales, like the Five-point Mean Opinion Score (MOS) scale (1 indicates lowest quality and 5 highest quality). Also, in order to provide a measure of the dispersion of QoE among users, the standard deviation σ can be used. Based on the aforesaid [38,39], proposed a QoE Fairness index which considers the lower bound L and the higher bound H of the rating scale. The formula is $F = 1 - \frac{2\sigma}{H-L}$. The QoE fairness index F value is bounded in the interval $[0,1]$ with 1 indicating the absolute QoE fairness (all users experience the same quality) and 0 indicating complete QoE unfairness.

In our investigation, for calculating the QoE fairness metric, the same formula is used. Here we assume as H the highest data rate and L as the lowest data rate that a D2D device can achieve in the D2D Network. The standard deviation σ is calculated by considering the Data Rate of each device in the network. The reason for selecting the aforesaid formula for calculating the QoE fairness metric is that the unit of measurement does not matter. Also the QoE fairness index F has some desired properties, like scale and metric independence (i.e., any linear transformation of the QoE values does not change the value of the fairness index).

7.1.3.4 Min and Max Percentage Changes in SE and PC

In order to calculate the min and max percentage changes of each investigated approach in SE and PC, the following calculations are used:

The minimum percentage change of SE of each approach is calculated using the Eq. 7.20.

$$S\underset{\mathrm{SE}}{\min}(\mathrm{UEs},\mathrm{app}) = \frac{f_{\mathrm{se}}(\mathrm{UEs},\mathrm{TP},\mathrm{app}) - \min \mathrm{SE} f_{\mathrm{ue}}(\mathrm{UEs})}{f_{\mathrm{se}}(\mathrm{UEs},\mathrm{TP},\mathrm{app})} \times 100 \tag{7.20}$$

$$\min \mathrm{SE} f_{\mathrm{ue}}(\mathrm{UEs}) = \min_{\mathrm{app}=\mathrm{UE},\ldots,\mathrm{MEC}} (f_{\mathrm{se}}(\mathrm{UEs},\mathrm{TP},\mathrm{app}))$$

The maximum percentage change of PC of each approach is calculated using the Eq. 7.21.

$$P\underset{\mathrm{PC}}{\max}(\mathrm{UEs},\mathrm{app}) = \frac{\max \mathrm{PC} F_{\mathrm{ue}}(\mathrm{UEs}) - f_{\mathrm{pc}}(\mathrm{UEs},\mathrm{TP},\mathrm{app})}{\max \mathrm{PC} F_{\mathrm{ue}}(\mathrm{UEs})} \times 100 \tag{7.21}$$

$$\max \mathrm{PC} F_{\mathrm{ue}}(\mathrm{UEs}) = \max_{\mathrm{app}=\mathrm{UE},\ldots,\mathrm{MEC}} (f_{\mathrm{pc}}(\mathrm{UEs},\mathrm{TP},\mathrm{app}))$$

For the simulation running instances TP was selected equal to 160 mW.

7.1.4 PERFORMANCE EVALUATION OBJECTIVES

In this section, we outline the performance evaluation objectives. Starting, with the aim to evaluate and compare of Sum Rate approach as shown in Section 6.2.1, Distributed Random, non-D2D UE with the initial instance (non-enhanced) of DAIS. Then, we aim to evaluate and compare the enhanced DSR and DAIS using the simulation environment described above, and also compare with the competing approaches described earlier. Fuzzy ART, DBSCAN, G-MEANS, and MEC are centralized unsupervised learning AI/ML clustering techniques that separate UEs into clusters, hence implementing ultra-dense networks, under the BS. It is worth noting that for the Cluster Head (CH) selection and the formation of the clusters, a heuristic algorithm was implemented (see Algorithm 6.5).

Furthermore, the Distributed Random (DR) approach (see Section 6.4) and the case where D2D communication is not used (non-D2D UE), are also included in the comparison. Table 7.3 shows each approach with the type of control and network knowledge that it needs.

Our simulative evaluation investigates the efficiency of each approach in terms of SE and PC during D2D communication. For this investigation, we simulated scenarios with different number of UEs and representative results related to scenarios with 50, 200, 500, and 1000 UEs are demonstrated and compared. Also, due to the high bandwidth requirements of 5G we set a target to offer a minimum sum rate of around 600 bits/s/Hz to all devices (e.g., around 12 bits/s/Hz per UE in a scenario with 50 UEs).

Additionally in our analysis, we examine the mean time (μ) of execution of each approach per UE, in terms of the time needed for the selection of the transmission mode.[5] The formula used is:

$$\mu = \frac{\sum_{1}^{N} TM'\text{selection}_{\text{time}}}{N}$$

However, depending on the type of control performed (i.e., Centralized, Semi-distributed, Distributed or DAI) by the approach, the conclusion time differs. More specifically:

In the case the approach uses centralized (i.e., FuzzyART, DBSCAN, GMEANS, MEC) or semi-distributed control, the conclusion is achieved when the transmission mode is selected for all D2D devices in the Network.

In the case the approach uses Distributed or DAI control (i.e., DSR, DAIS, DR) the conclusion is achieved when the transmission mode is selected for the specific D2D device.

Table 7.3
Investigated Approaches: Type of Control & Network Knowledge Needed

Approach(es) Investigated	Type of Control and Network Knowledge
DAIS	DAI (Distributed, decentralized with local knowledge)
DSR	Distributed with global knowledge
Distributed Random (DR)	Distributed with global knowledge
Fuzzy ART, DBSCAN, G-MEANS, MEC, and Centralized control with global knowledge	

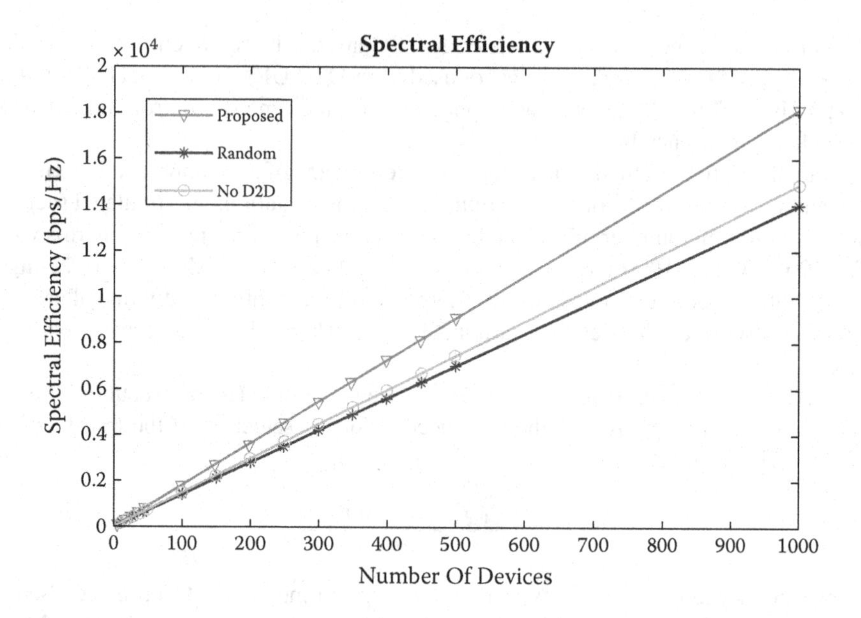

Figure 7.1 Spectral efficiency of different transmission modes.

7.1.5 PERFORMANCE EVALUATION RESULTS ON THE INITIAL DAIS AND SUM RATE PLANS

The performance related to the efficiency of each approach, in terms of SE and power, is evaluated using scenarios starting at 10 up to 1000 UEs in steps of 1 UE, using a mix of D2D devices and non D2D devices, dependant on the approach. Firstly, we examine the SE of DAIS. Figure 7.1 shows that our proposed solution has a better performance compared to a random clustering solution when no-D2D communication is used. The realized benefits are in the order of 30%. The most interesting result is that random clustering results in SE are even worse than direct UE-BS communication. Secondly, considering the power as shown in Figure 7.2 needed to realize the communication of the nodes, it is not surprising to see that clustering indeed requires less power. However, the proposed solution still outperforms the second best (i.e., no-D2D UE) by about 25%.

Within the proposed framework we have the ability to easily interchange metrics and parameters. In Section 6.1.2 we have argued on the feasibility of using WDR instead of Sum-Rate in our calculations. Figure 7.3 shows that the use of WDR does not reduce the SE of the system. The same happens if we consider an option in which an UE participates in the D2D communication depending on the remaining battery it has. Figure 7.4 shows no difference in SE.

On the contrary, by utilizing a battery threshold we are slightly increasing the required power for the communication, as evident by the slight differences shown in Figure 7.5.

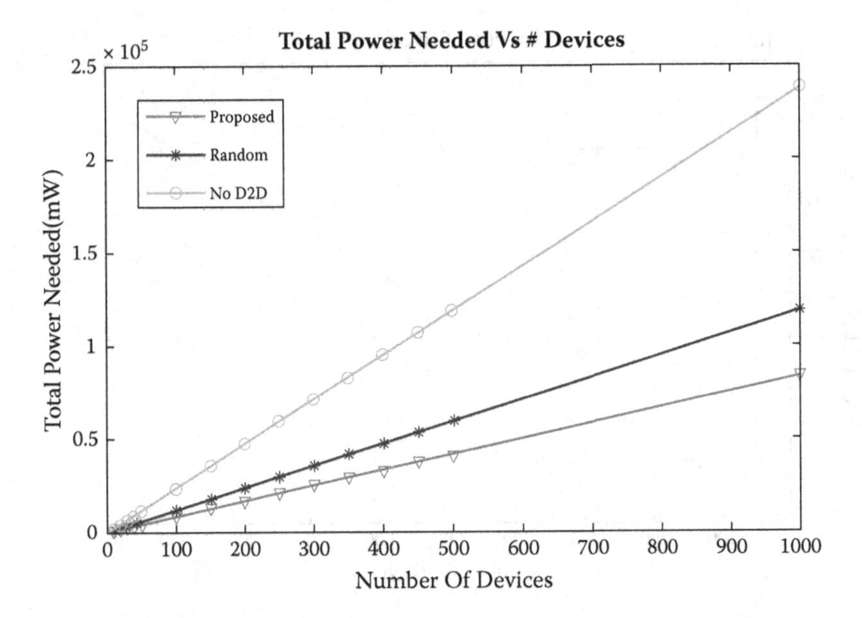

Figure 7.2 Power savings of different transmission modes.

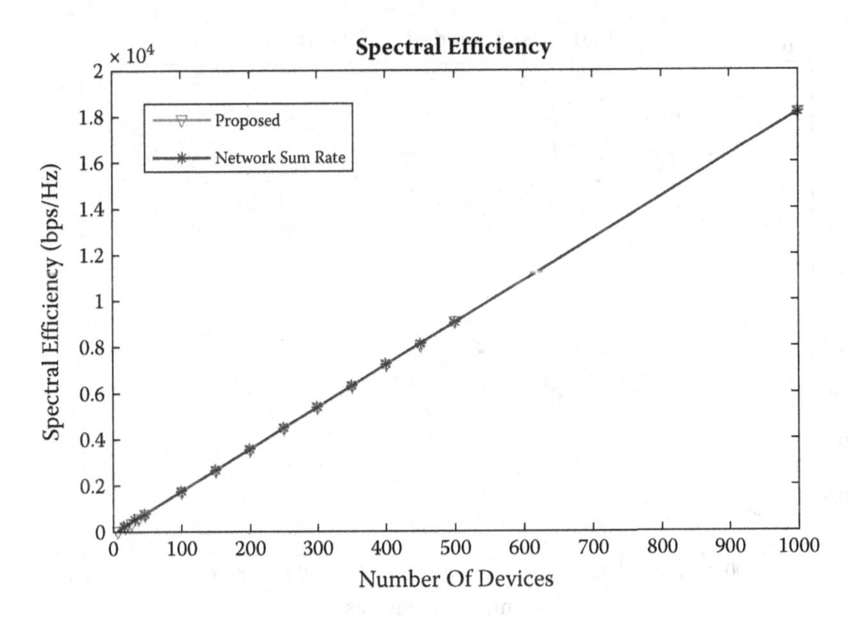

Figure 7.3 Spectral efficiency of different rate options.

A significant result, which validates our choice of WDR is that the computational time needed to perform sum-rate calculations is up to five (5) times greater than the

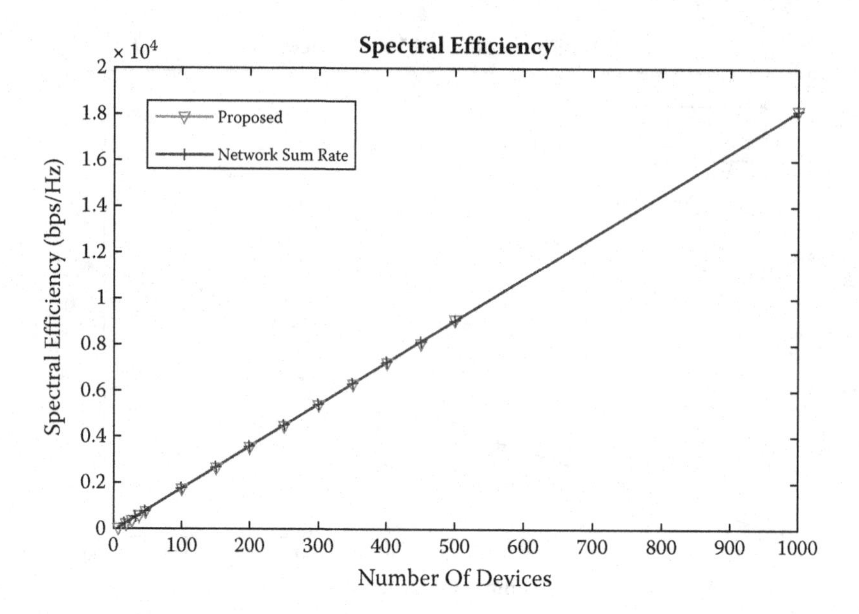

Figure 7.4 Spectral efficiency of different power options.

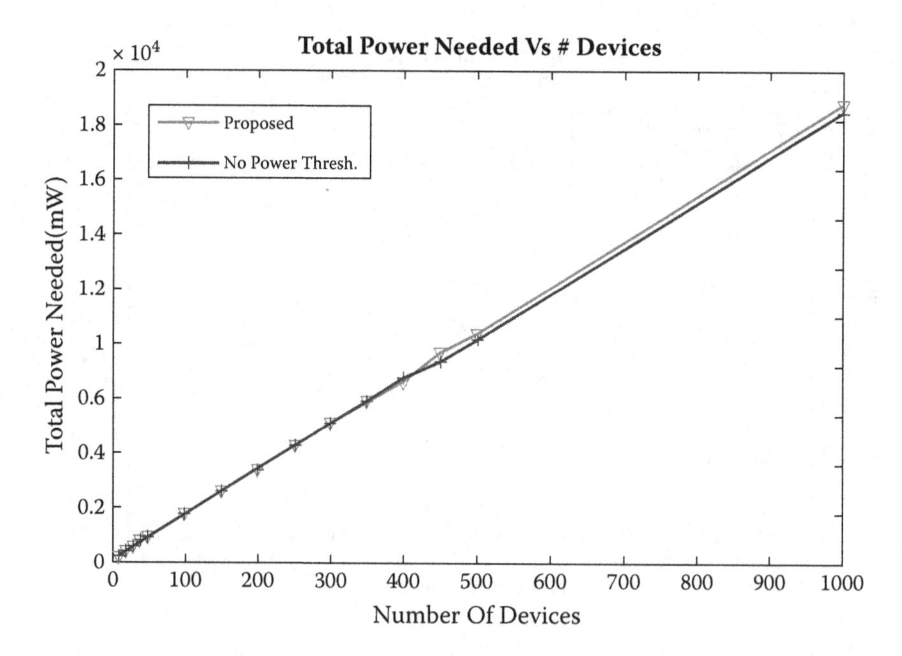

Figure 7.5 Power saved.

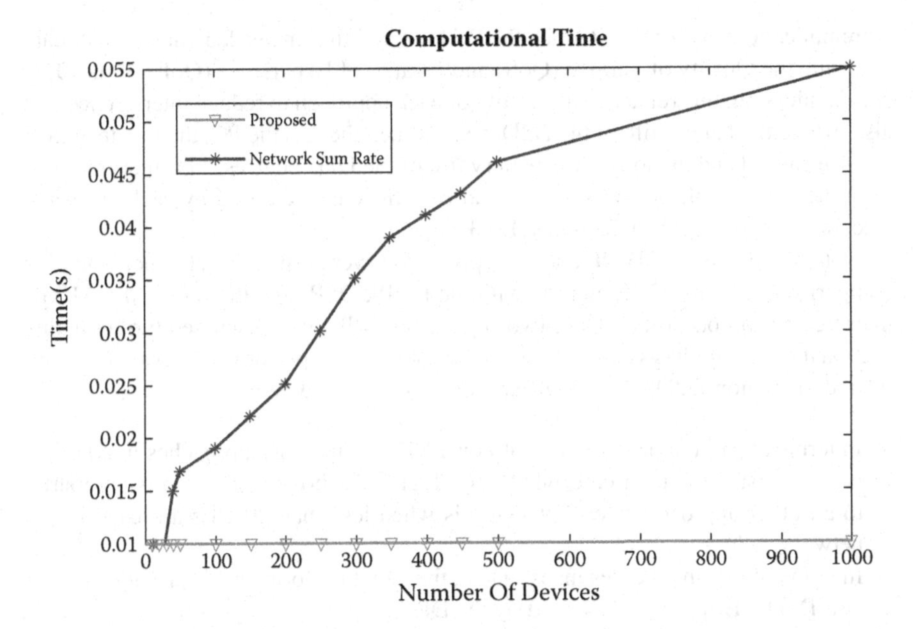

Figure 7.6 Computational complexity.

constant computation needed when we perform WDR calculations locally. This is ascribed to the fact that sum-rate needs to check all links in the network every time it needs to decide the transmission mode of a UE. As the number of UEs increases the computational time increases as well. In our case, the time to form a cluster is 100 ms for any device density, because the D2D UEs have all their link rates precalculated, so that WDR for the new connection is easily computed. Based on the evaluation results it became obvious that enhancements of DAIS are required. These enhancements include: (i) an additional parameter (number of D2D devices in proximity) to be considered in the transmission mode selection and (ii) the dynamically settable Weighted Data Rate (WDR) and Battery Power Level (BPL) thresholds, adapted and fine tuned for scenarios with a range from 10 to 1000 UEs. The performance of the enhanced DAIS is presented next (Figure 7.6).

7.1.6 PERFORMANCE EVALUATION RESULTS ON ENHANCED DAIS, DSR, AND UNSUPERVISED LEARNING CLUSTERING TECHNIQUES

The performance related to the efficiency of each approach, in terms of SE and PC, is evaluated using scenarios starting at 10 up to 1000 UEs in steps of 1 UE, using a mix of D2D devices and non D2D devices, dependant on the approach. In the results, we focused on 50, 200, 500, and 1000 UEs in our discussion, as indicative of the ranges 10–50, 50–200, 200–500, and 500–1000 UEs. In these scenarios we compared the ability of each approach: (i) To achieve high SE during D2D communication; this relates to the ability of each approach to provide higher Data Rates during D2D communication; and (ii) To Reduce the PC to the minimum needed but still adequate to ensure the Quality of Service (QoS) and Quality of Experience (QoE) of the D2D communication; this relates to the ability of each approach to reduce interference and also extend the battery life of the D2D devices (i.e., the less the PC, the less the interference caused and the longer the battery life of the D2D devices) . Furthermore, we examine the tradeoff between the SE and PC efficiency achieved by each approach (see Tables 7.4 and 7.5 in Section 7.1.6.3.)

Note that the non-D2D UE and DR approaches were used as a reference point for comparison in terms of SE and PC with the DAIS, DSR, and the rest of the AI/ML investigated approaches. As these two approaches will not be discussed further in this section the main findings extracted from the comparative performance evaluation and related to the non-D2D UE and DR approaches are stated here:

- In terms of SE, the performance of non-D2D UE and DR approaches in all cases is the worst. The main case where non-D2D UE shows good results compared to all other approaches (except DSR) is when less than 20 UEs are used in the network.
- In terms of PC, in all cases investigated, the worst performance is provided by the non-D2D UE approach followed by the DR.
- In terms of execution time (i.e., control decision delay), the DR provides the second best results, for all running instances.

The performed evaluations and the sections they appear are outlined below:

Compare the ability of each approach: (i) to increase the data rates (i.e., ability to increase the SE achieved); and (ii) to reserve power for the D2D devices (i.e., ability to reduce the PC to the minimum needed but still adequate to guarantee the Quality of Service (QoS) and Quality of Experience (QoE) of the communication). In particular:

- Section 7.1.6.1 examines the effect of TP on SE efficiency.
  Section 7.1.6.2 examines the effect of TP on the PC efficiency.
- Section 7.1.6.3 examines the effect of TP on SE and PC efficiency together, noting any tradeoffs.

For these evaluation results (except those related to the non-D2D UE approach,[6]) a "brute force" investigation is executed with the TP values of the links decreasing from 160 mW down to 60 mW, in steps of 10 mW.

Examine the TP needed to achieve maximum SE and minimum PC (see Section 7.1.6.4).

Compare the performance (i.e., gains achieved in terms of PC, SE, and the new metrics introduced) of the enhanced DAIS and DSR with the other competing approaches (see Section 7.1.6.5).

Examine the efficiency of clusters formed (e.g., number of clusters created, number of devices not assigned in clusters, etc.) and number of Messages needed to be Exchanged (e.g., message overloading) for forming the clusters by each approach.

Table 7.4
Minimum PC Achieved by Each Approach (50/200/500/1000 UEs)

Number of Devices	**50**		**Number of Devices**	**200**	
Approach (PC ASC)	**Min. PC(mW)**	**SE (bits/s/Hz)**	**Approach (PC ASC)**	**Min. PC(mW)**	**SE (bits/s/Hz)**
DSR	**2113.66**	**694.21**	DSR	**7356.14**	**2961.69**
FUZZYART	2317.59	644.36	DAIS	8651.99	2743.29
DAIS	2340.78	621.74	FUZZYART	9000.98	2564.46
MEC	2486.61	612.86	MEC	9798.44	2416.47
GMEANS	2673.88	539.08	GMEANS	10570.84	2178.41
DBSCAN	2704.54	537.24	DR	10900.2	2001.95
DR	2710.26	517.92	DBSCAN	11061.37	2087.71
non-D2D UE	11815.76	641.12	non-D2D UE	47226.84	2534.53
Number of Devices	**500**		**Number of Devices**	**1000**	
Approach (PC ASC)	**Min. PC(mW)**	**SE (bits/s/Hz)**	**Approach (PC ASC)**	**Min. PC(mW)**	**SE (bits/s/Hz)**
DSR	**17794.2**	**7519.95**	DSR	**34909.7**	**15244.78**
DAIS	21166.51	7005.5	GMEANS	37445.86	14607.23
FUZZYART	21350.21	6721.63	DAIS	41712.59	14163.34
GMEANS	23472.12	6199.31	FUZZYART	42431.2	13555.81
MEC	24864.22	5894	MEC	51932.67	11245.21
DBSCAN	27314.4	5259.18	DR	54666.68	9957.91
DR	27407.9	5006.03	DBSCAN	54856.89	10516.18
non-D2D UE	118326.78	5325.52	non-D2D UE	237032.28	12656.24

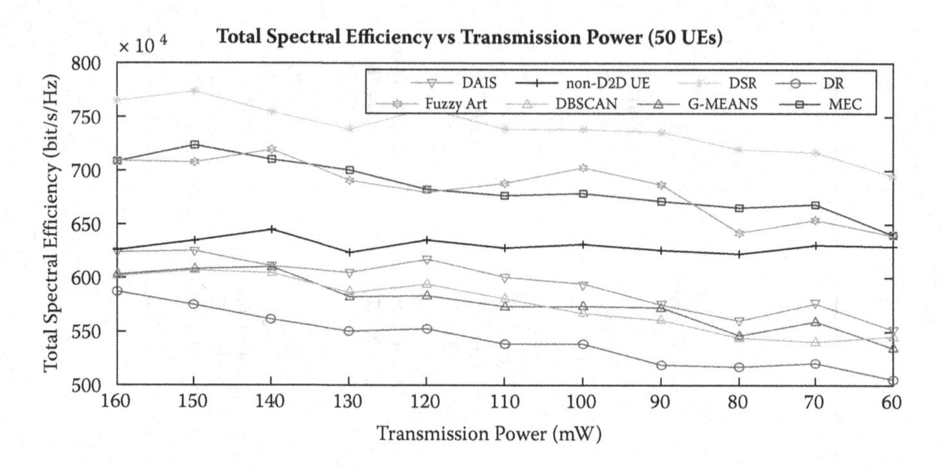

Figure 7.7 Total spectral efficiency vs link transmission power (10–50 UEs).

In this examination, we use 160 mW for the Transmission Power (TP) of the links Section (see Section 7.1.6.6).

Compare the QoE & QoS fairness among all approaches (see Section 7.1.6.7).

Compare each approach separately with the rest of the approaches in terms of SE, PC and mean time of execution (see Section 7.1.6.8).

7.1.6.1 SE Efficiency

In this evaluation scenario, we investigate the applicability of the investigated approaches to support a very large number of devices under the same network (mMTC) and at the same time to provide high service quality and quantity in order to achieve the users demanding bandwidth (eMBB) with the use of SE examination and a different number of devices (i.e., 50, 200, 500, and 1000) in the simulation. The purpose of the scenario is to examine the achievement of the two use cases (i.e., mMTC, eMBB) in the 5G use case triangle [40].

The results related to the SE achieved by each approach are illustrated in Figures 7.7–7.11.

From the results collected we can identify the best performing approaches in terms of SE: (i) For scenarios with 10–50 UEs the DSR and FuzzyART followed by MEC and DAIS (with a small difference among them); (ii) For scenarios with 50–500 UEs the DSR and DAIS followed by FuzzyART; and (iii) For scenarios with 500–1000 UEs the DSR and GMEANS followed by DAIS. By best performance, we mean the selection of the transmission mode that will increase the SE in the highest achievable value and reduce the PC in the lowest achievable value.

An approach that can have full knowledge of the existing network structure (i.e., the UEs with their associated links), is expected to achieve the most appropriate selection of the best transmission mode and accomplish the best results in terms of SE and PC. As expected the enhanced DSR provides the best results since it

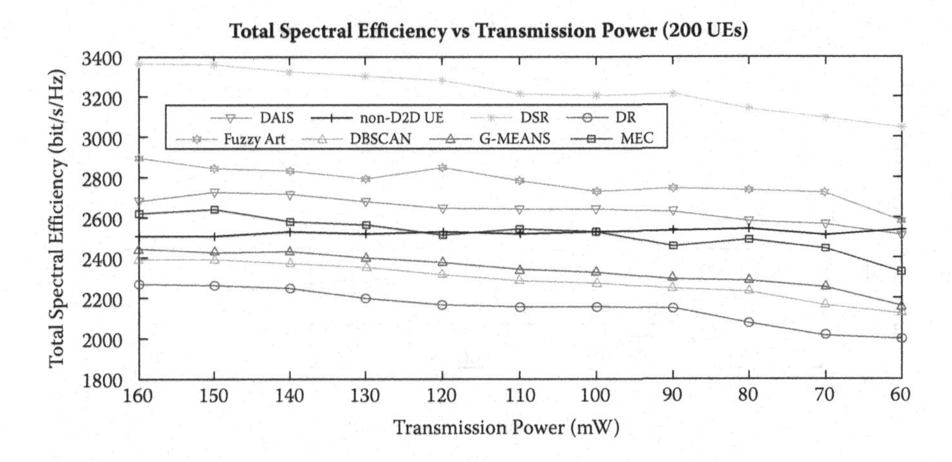

Figure 7.8 Total spectral efficiency vs link transmission power (51–200 UEs).

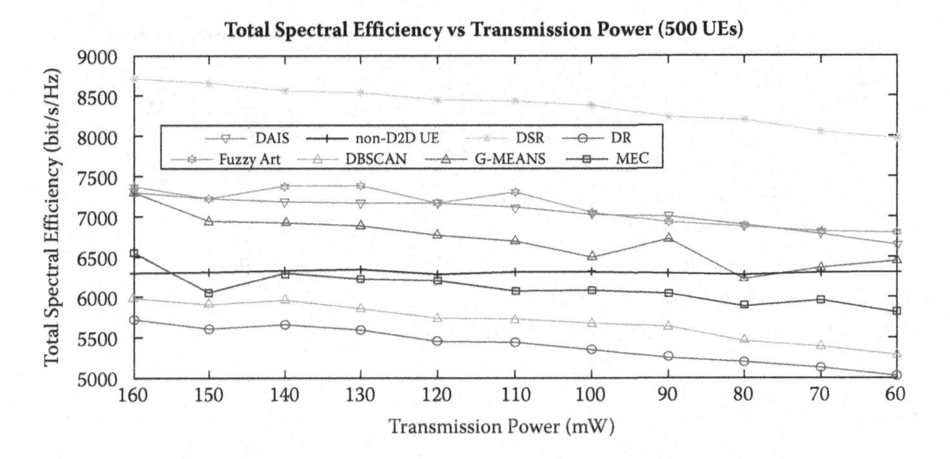

Figure 7.9 Total spectral efficiency vs link transmission power (201–500 UEs).

is the only approach which selects the transmission mode by having full network knowledge. Note that DAIS remains among the top three list, considering the range from 50 to 1000 UEs, that can achieve high SE and still achieve the SINR required at the Receiver for preserving the fidelity of the signal and achieve the requested QoS.

7.1.6.2 PC Efficiency

In this evaluation scenario, we examine the energy reservation of the investigated approaches with the use of PC. In order to achieve a reduction in energy consumption which is a 5G requirement for utilization of green energy (solar panels) [41].

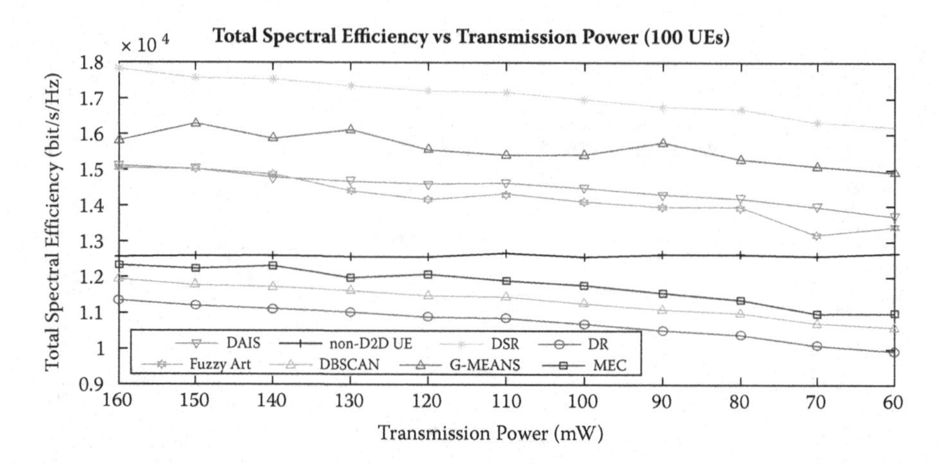

Figure 7.10 Total spectral efficiency vs link transmission power (501–1000 UEs).

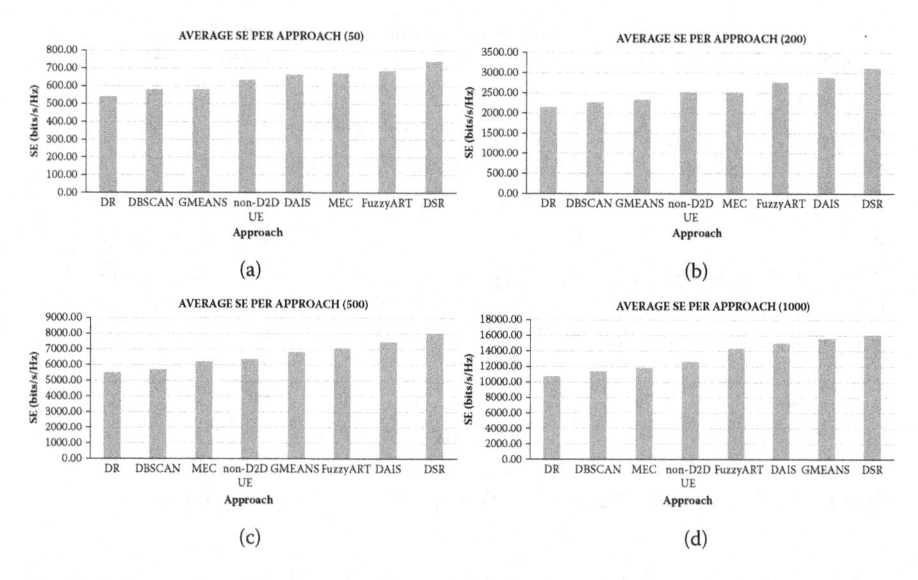

Figure 7.11 Average SE achieved by each approach: (a) 50 UEs, (b) 200 UEs, (c) 500 UEs, and (d) 1000 UEs.

The results related to the PC efficiency achieved by each approach (for the simulated scenarios) are illustrated in Figures 7.12–7.16.

From the results collected we can identify the best performing approaches in terms of PC: (i) For scenarios with 10–50 UEs, the DSR and FuzzyART followed by MEC and DAIS (with a small difference among them); (ii) For scenarios with 50–500 UEs the DSR and DAIS followed by FuzzyART; (iii) For scenarios with 500–1000 UEs the DSR and GMEANS followed by DAIS. As expected given the full knowledge of

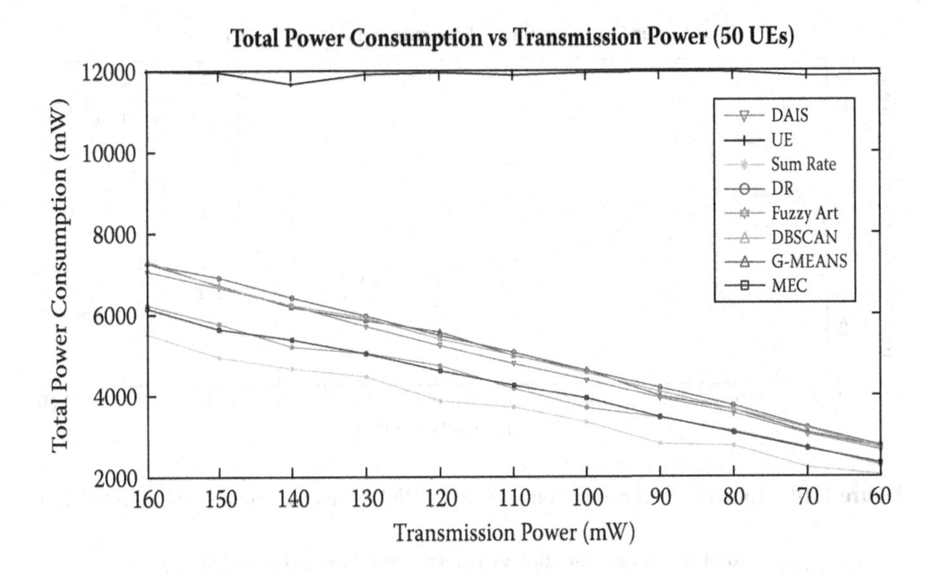

Figure 7.12 Total power consumption achieved vs link transmission power (1–50 UEs).

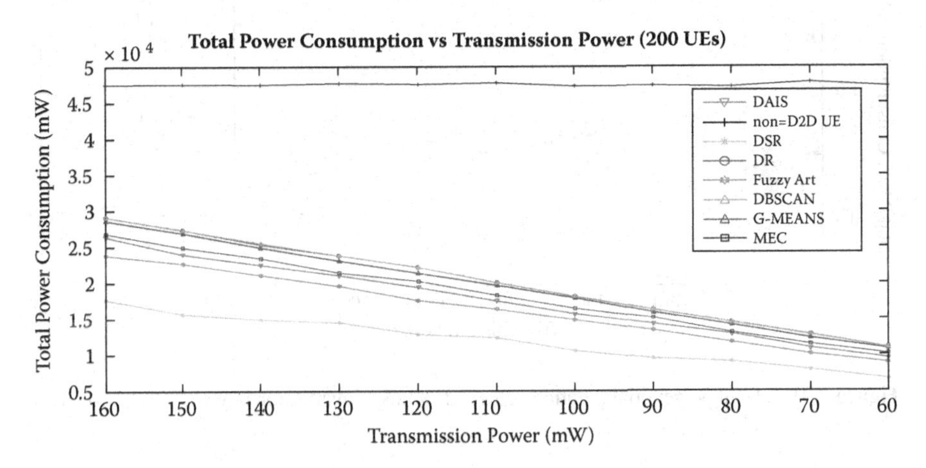

Figure 7.13 Total power consumption achieved vs link transmission power (51–200 UEs).

enhanced DSR, it outperfoms all. Worth noting that considering the range from 50 to 1000 UEs, DAIS remains within the top 3 that can achieve low PC and still ensure the QoS of the communication.

7.1.6.3 SE and PC Efficiency Tradeoff

In order to achieve increased SINR at the Receiver, and perhaps preserve the fidelity of the signal and its SE efficiency, it is expected that an increase in the TP of the

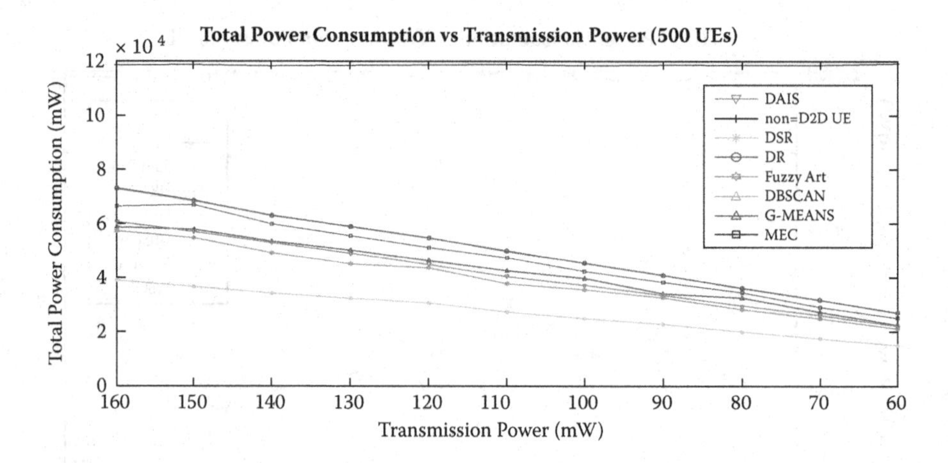

Figure 7.14 Total power consumption achieved vs link transmission power (201–500 UEs).

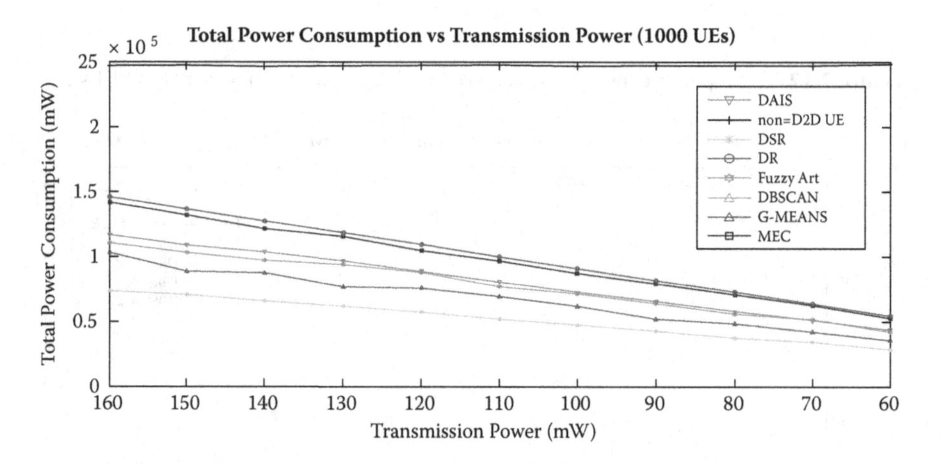

Figure 7.15 Total power consumption achieved vs link transmission power (501–1000 UEs).

links would normally be required. However, this would result in reduced PC efficiency. For the same reason, reducing the Bandwidth Efficiency expectation (i.e., reducing SE), one can expect a decrease in the total PC and thus an increase in the PC efficiency. Thus, in this evaluation scenario, we examine the trade-off between PC and SE in each of the investigated approaches targeting the identification of the most appropriate approach for the achievement of the minimization of PC with the least reduction in SE by changing the TP in a different number of devices running instance, targeting the achievement of reducing energy consumption, which is a 5G requirement. Indicative results of this trade-off appears in Figure 7.17.

Below, we discuss a number of observations regarding all approaches, such as: (i) the diminishing improvements in SE if one increases Link TP, and hence PC (see

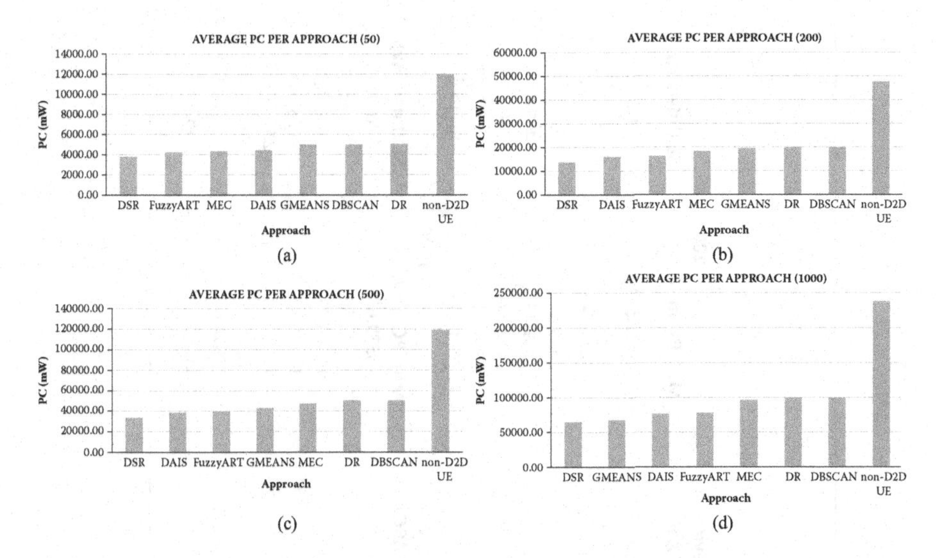

Figure 7.16 Average PC achieved by each approach: (a) 50 UEs, (b) 200 UEs, (c) 500 UEs, and (d) 1000 UEs.

Figure 7.17); and (ii) the effect of an increased number of UEs (50/200/500/1000 UEs), as shown in Figures 7.11, 7.16, and 7.17, along with equations B.3 and B.4. Thus one has to consider carefully the gains in SE versus the loss in PC efficiency. Results related to this tradeoff are included in Table 7.4 (showing the minimum PC achieved by each approach) and Table 7.5 (showing the maximum SE achieved by each approach), listed in ascending order for PC and descending order for SE.

From these results, the following observations are made:

i) For the scenarios with 10–50 UEs, the best improvement in terms of PC is achieved by DAIS (i.e., 63.43% improvement; see Table 7.4 and Figure 7.12). In order to achieve the aforesaid PC improvement, the SE achieved by DAIS has, as a tradeoff, a negligible decrease of 8.56% (see Figure 7.7). On the other hand, the best performance in terms of SE, is provided by DSR. More specifically, DSR provided the least negative effect on SE (i.e., only 9.34% reduction; see Table 7.5 and Figure 7.7), while targeting increased PC efficiency (i.e., a gain of 57.56% reduction on the total PC is achieved; see Figure 7.12). Regarding the maximum negative effect on SE, it is provided by MEC (i.e., 12.35% reduction; see Tables 7.5 and 7.7) which however has, as a tradeoff, a gain of 57.63% reduction on the total PC achieved (see Figure 7.12).
ii) For the scenarios with 50–200 UEs, the best improvement in terms of PC is achieved by MEC (i.e., 62.94% improvement; see Table 7.4 and Figure 7.13). In order to achieve the aforesaid PC improvement, the SE achieved by MEC has, as a tradeoff, a negligible decrease, of 8.78% (see Figure 7.8). On the other hand, the best performance in terms of SE, is provided by DAIS. More specifically, DAIS provided the least negative effect on SE (i.e., only 8.82% reduction; see Table 7.5

Table 7.5
Maximum SE Achieved by each approach (50/200/500/1000 UEs)

Number of Devices	**50**		**Number of Devices**	**200**	
Approach (PC ASC)	**Min. PC(mW)**	**SE (bits/s/Hz)**	**Approach (PC ASC)**	**Min. PC(mW)**	**SE (bits/s/Hz)**
DSR	**765.8**	**4981.02**	DSR	**3271.5**	**19435.3**
FUZZYART	728.82	5146.45	DAIS	3008.91	21789.88
MEC	699.21	5868.82	FUZZYART	2914.76	18678
DAIS	694.53	6401.24	MEC	2649.34	26440.38
non-D2D UE	641.13	11815.76	non-D2D UE	2539.43	47742.4
GMEANS	610.16	6766.34	GMEANS	2483.77	28039.7
DBSCAN	604.42	6750.76	DBSCAN	2397.51	29272.23
DR	574.76	6870.46	DR	2271.97	29112.87
Number of Devices	**500**		**Number of Devices**	**1000**	
Approach (PC ASC)	**Min. PC(mW)**	**SE (bits/s/Hz)**	**Approach (PC ASC)**	**Min. PC(mW)**	**SE (bits/s/Hz)**
DSR	**8233.79**	**47709.14**	DSR	**16579.65**	**94670.61**
DAIS	7727.78	55739.03	GMEANS	16332.69	95643.39
FUZZYART	7235.15	50915.72	DAIS	15585.83	111766.76
GMEANS	7099.98	48331.03	FUZZYART	14926.12	112819.88
MEC	6453.19	67815.77	non-D2D UE	12656.24	237032.27
non-D2D UE	6330.51	118326.78	MEC	12531.8	120303.45
DBSCAN	5956.12	73342.62	DBSCAN	11974.08	146603.95
DR	5716.09	72803.35	DR	11333.15	146351.78

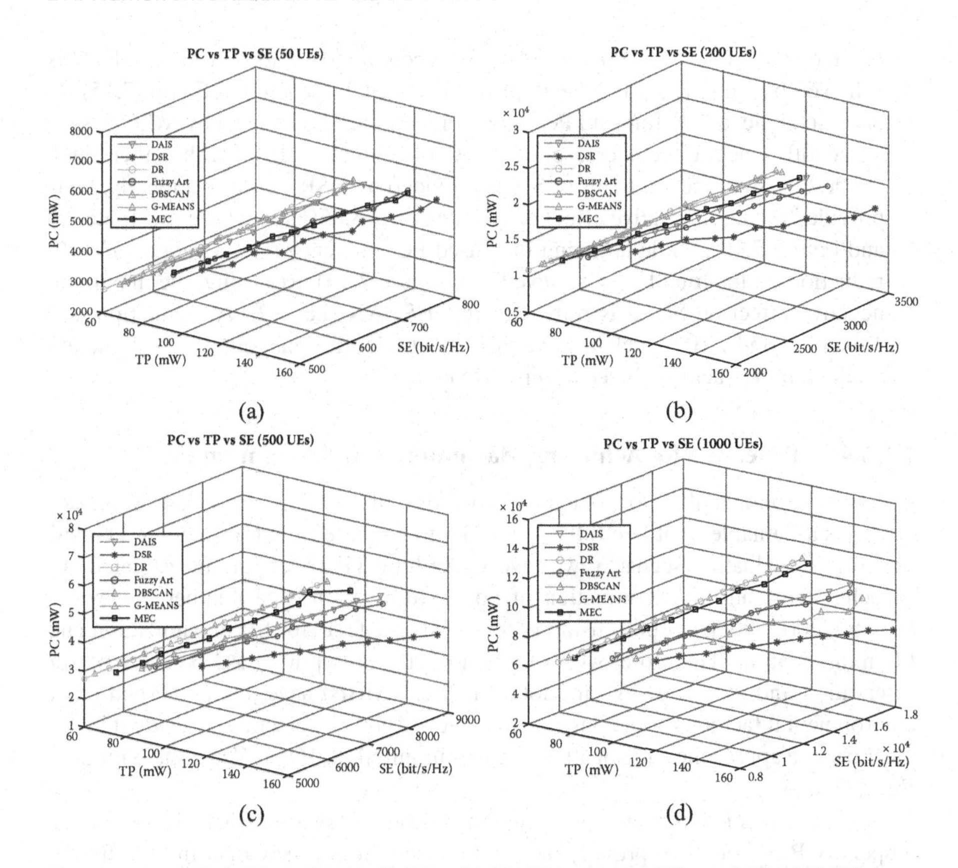

Figure 7.17 PC vs TP vs SE: (a) 50 UEs, (b) 200 UEs, (c) 500 UEs, and (d) 1000 UEs.

and Figure 7.8), while targeting increased PC efficiency (i.e., a gain of 62.52% reduction on the total PC is achieved; see Figure 7.13). Regarding the maximum negative effect on SE, it is provided by GMEANS (i.e., 12.92% reduction; see Tables 7.5 and 7.8) which however has, as a tradeoff, a gain of 62.30% reduction on the total PC achieved (see Figure 7.13).

iii) For the scenarios with 200–500 UEs, the best improvement in terms of PC is achieved by Fuzzy ART (i.e., 64.26% improvement; see Table 7.4 and Figure 7.14). In order to achieve the aforesaid PC improvement, the SE achieved by Fuzzy ART has, as a tradeoff, a negligible decrease of 6.32% (see Figure 7.9). On the other hand, the best performance in terms of SE, is provided again by Fuzzy ART (excluding non-D2D UE). More specifically, Fuzzy ART provided the least negative effect on SE (i.e., only 7.29% reduction; see Table 7.5 and Figure 7.9), while targeting increased PC efficiency (i.e., a gain of 50.69% reduction on the total PC is achieved; see Figure 7.14). Regarding the maximum negative effect on SE, it is provided by DR (i.e., 12.42% reduction; see Tables 7.5 and 7.9) which however has, as a tradeoff, a gain of 62.55% reduction on the total PC achieved (see Figure 7.14).

iv) For the scenarios with 500–1000 UEs, the best improvement in terms of PC is achieved by MEC (i.e., 63.67% improvement; see Table 7.4 and Figure 7.15). In order to achieve the aforesaid PC improvement, the SE achieved by MEC has, as a tradeoff, a negligible decrease, of 8.01% (see Figure 7.10). On the other hand, the best performance in terms of SE, is provided by DSR. More specifically, DSR provided the least negative effect on SE (i.e., only 8.05% reduction; see Table 7.5 and Figure 7.10), while targeting increased PC efficiency (i.e., a gain of 63.12% reduction on the total PC is achieved; see Figure 7.15). Regarding the maximum negative effect on SE, it is provided by DBSCAN (i.e., 12.17% reduction; see Tables 7.5 and 7.10) which however has, as a tradeoff, a gain of 62.58% reduction on the total PC achieved (see Figure 7.15).

7.1.6.4 TP Needed for Achieving Maximum SE and Minimum PC

Due to environmental factors, such as Path Loss, Shadowing, and Noise, some approaches are unable to utilize full TP in order to increase SE while keeping PC low. So, in this evaluation scenario, we investigate how TP affects the investigated approaches targeting the examination of power reservation by a number of devices (i.e., 50, 200, 500, and 1000) using specific values of TP, for SE maximization and PC minimization. Therefore, we examine the effects that total PC (i.e., for power reservation and green energy) and total SE (i.e., eMBB) have due to TP and other environmental factors, for the purpose to achieve 5G requirements. As evident in Table 7.6 some approaches do not attain the maximum SE at 160 mW and the minimum PC at 60 mW.

The aforementioned results prompted the power reservation Algorithm shown in Appendix B.1. There we provide the implementation of a new plan in DAI framework that can be executed by the BDIx agents targeting to decrease the TP with the least reduction of SE and the maximum gains in PC according to D2D device requirements (i.e., QoS). The plan is called "Distributed Artificial Intelligence Power Reservation (DAIPPR) Plan based on TP" and it can be activated when the Battery Power Level of a D2D-Relay Device drops below a threshold (i.e., 50%). However, as this is out of the scope of this section , this DAIPPR plan will be investigated further as future directions.

7.1.6.5 Evaluation of the D2D Effectiveness, Stability, and Productivity Metrics

In addition to the above results, we have evaluated the new metrics defined in Section 7.1.3, considering the effect of the number of D2D devices. In this evaluation scenario, we evaluate per approach the following: (i) how close the results of SE/PC are to the best SE/PC (PC/SE Effectiveness); (ii) the density and how the results are spread close to the mean of best SE/PC (PS/SE Stability); and (iii) the gain and loss of each approach by comparing the result from the previous step (SE/PC productivity). Consequently, with this scenario, we show how good each approach is in terms of SE/PC results, how close to the best results is, and how stable is.

Table 7.6
TP Needed for Achieving Maximum SE and Minimum PC (50/200/500/1000 UEs)

TP for Minimum PC per Approach				
# Devices	**50**	**200**	**500**	**1000**
Approach		Min PC TP		
DAIS	60	60	60	60
non-D2D UE	60	**140**	**100**	**90**
DSR	60	60	60	60
DR	60	60	60	60
FUZZYART	60	60	60	60
DBSCAN	60	60	60	60
GMEANS	60	60	60	60
MEC	60	60	60	60
TP for Maximum SE per Approach				
# Devices	**50**	**200**	**500**	**1000**
Approach		Max SE TP		
DAIS	160	**150**	160	160
non-D2D UE	**60**	**100**	**100**	**90**
DSR	**150**	160	160	160
DR	**150**	160	160	160
FUZZYART	**140**	**130**	**140**	160
DBSCAN	**150**	160	160	160
GMEANS	**150**	160	**130**	160
MEC	**150**	160	160	**140**

Firstly, we evaluate each approach based on their D2D SE Effectiveness (Eq. 7.6) and D2D PC Effectiveness (Eq. 7.7) and then jointly as D2D effectiveness (i.e., both SE and PC Effective). DSR is D2D effective, whereas DAIS, Fuzzy ART, MEC are only D2D SE Effective (Table 7.7). Likewise, we evaluate each approach based on their D2D SE Stability (Eq. 7.12), and D2D PC Stability (Eq. 7.13) and jointly as D2D Stability. Again, the DSR approach is the only D2D stable approach. DAIS, Fuzzy ART, MEC are only D2D SE Stable (Table 7.7). Regarding D2D SE Productivity (Eq. 7.16), D2D PC Productivity (Eq.7.19) and jointly as D2D Productivity, we can see that there is a direct relation between SE Productivity and PC Productivity (see Figures 7.18 and 7.19). DSR and DAIS approaches, both have gains in terms of SE/PC Productivity (more than 80%, basically around 85% as shown in Figure 7.18 and Figure 7.19). Thus, we can say that, compared to all other investigated approaches, DSR and DAIS are the only approaches that are D2D productive.

Table 7.7
AI/ML D2D Effectiveness and Stability

Approach	D2D Optimum Effectiveness in Percentage (≥80%)					D2D Stability in Percentage (≤5%)				
	S.E.	SE Eff.	P.C.	PC Eff.	Effective	S.E.	SE Stab.	P.C.	PC Stab.	Stable
DAIS	**83.0**	✓	71.3			**1.8**	✓	7		
non D2D UE	78.4		40.0			7.0		7		
DSR	**99.08**	✓	**99.8**	✓	✓	**0.8**	✓	**0.9**	✓	✓
DR	71.3		65.0			7.2		12.2		
Fuzzy ART	**88.0**	✓	78.0			4.7	✓	10		
DBSCAN	75.0		66.0			8.5		13.1		
G-MEANS	79.0		69.0			5.5		9.3		
MEC	**83.0**	✓	72.7			10.3		15.5		

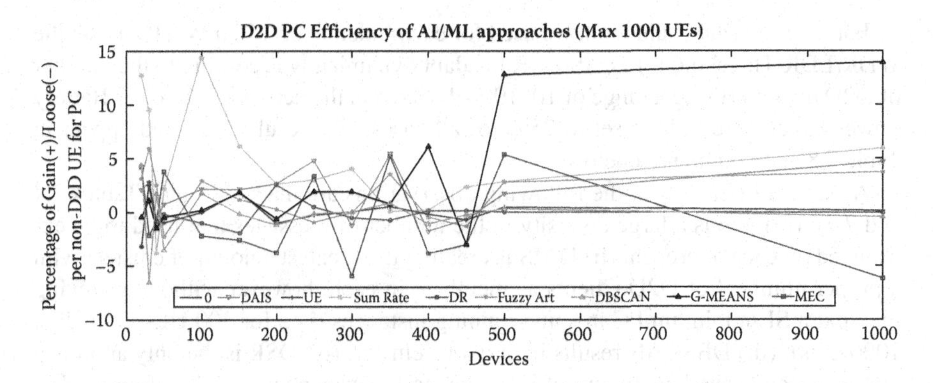

Figure 7.18 PC Gain and loss of different approaches by comparing result from previous step.

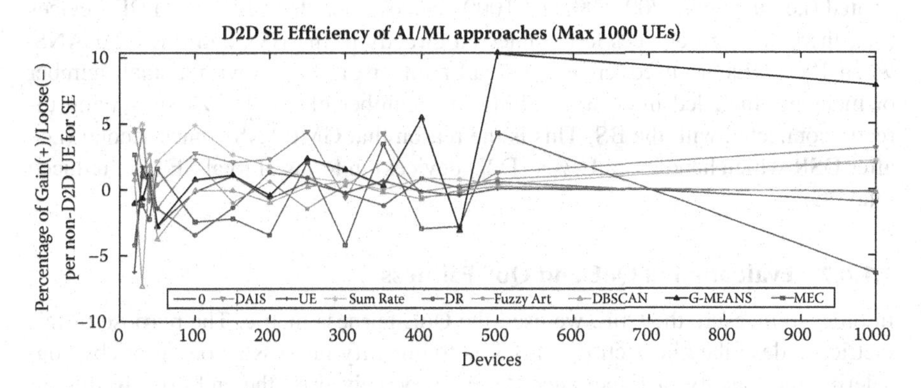

Figure 7.19 SE gain and loss of different approaches by comparing result from previous step.

7.1.6.6 Evaluation of Cluster Formation, Message Exchange, and Control Decision Delay

In this section, we evaluate each approach in terms of the characteristics of the created clusters. These include the maximum number of clusters created and the number of UEs (i.e., D2D devices) per cluster (maximum density of cluster). Also, the number of UEs (i.e., non D2D devices) that remain directly connected with the BS as mobile network devices without selecting transmission mode, is measured. Additionally, for each approach, we examine the number of messages used for selecting the transmission mode for all D2D devices. Note that this is a key factor that is highly associated with the time needed (i.e., the more the messages needed to be exchanged, the more the delay) by each approach for control decision-making. Therefore, we investigate how clusters are formed in terms of best position[7] by the investigated approaches, number of messages exchanged to conclude and how fast it is in terms of execution.

For the aforesaid evaluation, we used for each link a fixed 160 mW TP and set the WDR/LDR Threshold for DAIS/DSR to adapt dynamically according to the number of D2D devices (using a range of 10–1000 devices) in the network.[8] Also the Battery Power Level Threshold is set to 75% for all cases. The results collected appear in Table 7.8 and commented below.

As shown in Table 7.8, the following observations are made (see also Tables 7.4 and 7.5): (i) there is a large diversity in the number of messages that need to be exchanged by each approach; (ii) DAIS is creating the greatest amount of clusters with a proper number[9] of D2D Clients as members in each, however without achieving maximum SE/minimum PC in some running instances (i.e., for 200 UEs, 500 UEs, 1000 UEs); (iii) DBSCAN results in only one cluster; (iv) DSR is the only approach that needs an excessive amount of messages to be exchanged and therefore it takes a lot of time to conclude and decide the transmission mode of the D2D devices; (v) the DAIS, DSR and DR are the only approaches that, in all running instances investigated (i.e., using 50, 200, 500, and 1000 devices), handles all UEs as D2D devices (i.e., there are zero devices left connected directly to the BS); and (vi) GMEANS, when 1000 UEs are used, creates a small number of clusters with a small number of members included in each, resulting in a number of non D2D UEs staying directly connected with the BS. This is the reason that GMEANS comes second (i.e., after DSR which handles all UEs as D2D devices) in terms of total SE (and reduced total PC).

7.1.6.7 Evaluation of QoE and QoS Fairness

In order to measure the QoE, we used the QoE fairness metric. The purpose of this metric, as described in Section 7.1.3.3, is to quantify fairness among users by considering the Quality of Experience (QoE) as perceived by the end user. In this investigation, the following simulation parameters and constrains were set: (i) the TP is set to 160 mW; (ii) the higher bound (H) in the scale of fairness is the maximum achievable data rate[10] by a UE in the network in the same running instance; (iii) the lower bound (L) in the scale of the fairness investigation is the minimum achievable data rate achieved by a UE in the network in the same running instance. Hence, in this evaluation scenario, we investigate how fair in terms of QoE and QoS are the investigated approaches. Note that QoE and QoS are always requirements in network communication.

Therefore, the L and H are set in order to check how fair the investigated approaches are among all users. The rest of the simulation parameters and constraints are the same as in the previous investigations. In this section, we examine the QoE fairness of our approaches in conjunction with the non-D2D UE approach in terms of network utilization.

As can be seen from Table 7.9, all the investigated approaches are QoE fair in terms of network usage (see Section 7.1.3.3 on how QoE fairness is measured). This is indicated by their QoE fairness values (e.g. >60%), which are very close to the QoE fairness value achieved by the non-D2D UE approach. Note that the non-D2D UE approach is considered to be the fairest approach in terms of data rate due to the

Table 7.8
Clusters and Messages for 50, 200, 500 and 1000 Devices

Approach	Maximum Number of Devices per Cluster	Devices Remaining Connected to BS (non D2D UEs)	Number of Messages Exchanged	Number of Clusters Created	Number of Devices Used (Running Instance)
DAIS	6	0	65	**13**	**50**
Non-D2D UE	0	50	50	0	**50**
DSR	6	0	**1336**	12	**50**
DR	12	0	2	3	**50**
FuzzyART	6	14	144	8	**50**
DBSCAN	12	38	74	1	**50**
GMEANS	**18**	32	75	1	**50**
MEC	9	22	104	5	**50**
DAIS	**146**	0	230	**26**	**200**
Non-D2D UE	0	200	200	0	**200**
DSR	26	0	**20,321**	25	**200**
DR	21	0	2	7	**200**
FuzzyART	34	43	595	7	**200**
DBSCAN	50	150	300	1	**200**
GMEANS	49	128	344	3	**200**
MEC	38	92	414	4	**200**

(Continued)

Table 7.8 (*Continued*)
Clusters and Messages for 50, 200, 500 and 1000 Devices

Approach	Maximum Number of Devices per Cluster	Devices Remaining Connected to BS (non D2D UEs)	Number of Messages Exchanged	Number of Clusters Created	Number of Devices Used (Running Instance)
DAIS	64	0	556	**60**	**500**
Non-D2D UE	0	500	500	0	**500**
DSR	39	0	**125,790**	42	**500**
DR	26	0	3	7	**500**
FuzzyART	**70**	138	1493	8	**500**
DBSCAN	**120**	380	737	1	**500**
GMEANS	68	265	967	5	**500**
MEC	71	331	836	4	**500**
DAIS	173	0	1058	**60**	**1000**
Non-D2D UE	0	1000	1000	0	**1000**
DSR	110	0	**501,561**	63	**1000**
DR	52	0	5	8	**1000**
FuzzyART	112	460	2994	8	**1000**
DBSCAN	**220**	780	1438	1	**1000**
GMEANS	87	66	2858	13	**1000**
MEC	**224**	630	1739	4	**1000**

Table 7.9
QoE Fairness of each Approach

Number of Devices	DAIS	non-D2D UE	DSR	DR	FuzzyART	DBSCAN	GMeans	MEC
50	0.62	0.66	0.61	0.60	**0.94**	0.64	0.64	0.59
100	0.65	0.65	**0.70**	0.66	**0.95**	0.64	0.68	0.65
200	0.68	0.65	0.66	0.66	**0.93**	**0.70**	0.65	0.68
500	**0.71**	**0.73**	**0.70**	**0.73**	**0.94**	**0.76**	**0.70**	**0.71**
750	**0.76**	**0.75**	0.74	**0.75**	**0.92**	**0.73**	**0.76**	**0.72**
1000	**0.78**	**0.71**	**0.77**	**0.82**	**0.94**	**0.77**	**0.73**	**0.74**

frequency allocation of a single dedicated channel to each UE with a pre-specified data rate. The important observation from the table is that FuzzyART is QoE fair with the highest score followed by DR and DAIS. Additionally, we can see that QoE fairness values higher than 70% are achieved when we have a large number of devices (i.e., $\geq$500). The reason is that the network is more dense, due to the high number of devices, resulting in clusters with more members and more efficient and back-hauling links (Table 7.10).

Regarding Jain's fairness index for QoS, DAIS with non-D2D UE and then DSR along with Fuzzy ART are the QoS fairest. This QoS result was expected since DAIS, DSR and Fuzzy ART increase their SE (and reduce the PC) with the entering of new UEs in the D2D network. So, even though both DAIS and DSR offer autonomicity and distributed control at each UE, at the same time they also assure QoS fairness compared to other centralized approaches (e.g., non-D2D UE, Fuzzy ART).

Table 7.10
QoS Fairness of each Approach

Number of Devices	DAIS	non-D2D UE	DSR	DR	FuzzyART	DBSCAN	GMeans	MEC
50	**0.95**	**0.96**	**0.93**	**0.94**	**0.94**	0.63	0.67	0.63
100	**0.93**	**0.96**	**0.93**	**0.95**	**0.93**	0.69	0.64	0.72
200	**0.95**	**0.96**	**0.96**	**0.95**	**0.93**	0.70	0.68	0.67
500	**0.95**	**0.95**	**0.94**	**0.93**	**0.92**	0.80	0.75	0.78
750	**0.96**	**0.95**	**0.95**	**0.92**	**0.94**	0.75	0.73	0.73
1000	**0.95**	**0.95**	**0.95**	**0.92**	**0.93**	0.77	0.75	0.81

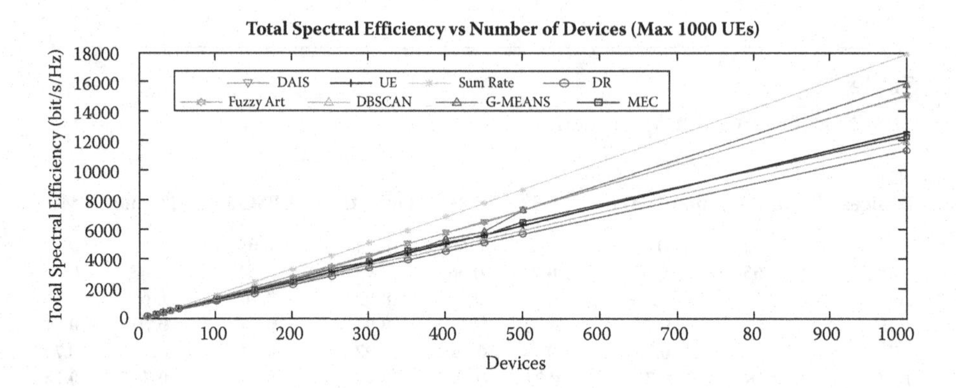

Figure 7.20 Total spectral efficiency vs number of devices of different approaches.

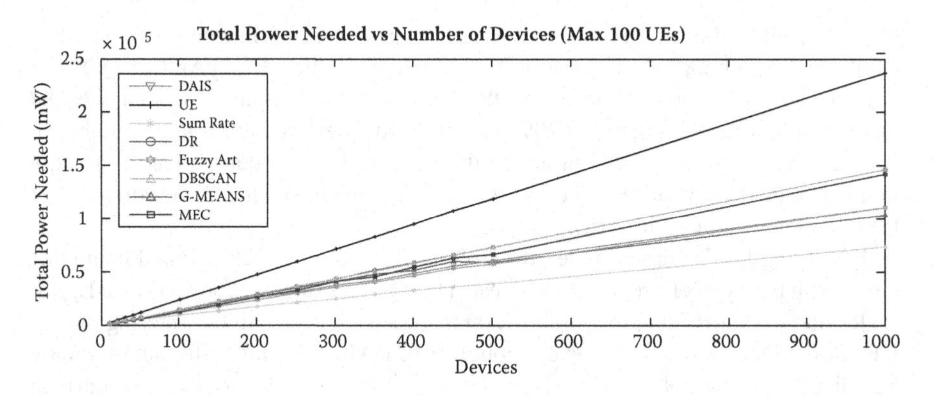

Figure 7.21 Power needed vs number of devices of different approaches.

7.1.6.8 Comparative Evaluation of Each Investigated Approach

The performance of each approach is compared with all other investigated approaches in terms of total SE (i.e., Sum Rate) achieved (see Figure 7.20), total PC needed (see Figure 7.21) and total time needed for finalizing execution (see Figure 7.22). For this comparison, a predefined link TP of 160 mW is used for all approaches.

Starting with DAIS approach, in terms of SE, benefits are provided when 500 UEs are used, reaching performance close to DSR. However, it underperforms compared to the other approaches for a network with a small number of devices (i.e., 20 UEs). On the other hand, from 50 UEs and above, DAIS is better than the DR, DBSCAN, G-MEANS and non-D2D-UE approaches. Furthermore, with 1000 UEs (maximum number of UEs examined), it ranks third. In terms of PC, with 200–500 UEs, DAIS is better than DBSCAN, MEC, G-MEANS, non-D2D-UE approach, and DR and has the same PC with Fuzzy ART. With 1000 UEs, DAIS ranks third with Fuzzy ART. Regarding the total time needed for finalizing execution, DAIS is the fastest approach.

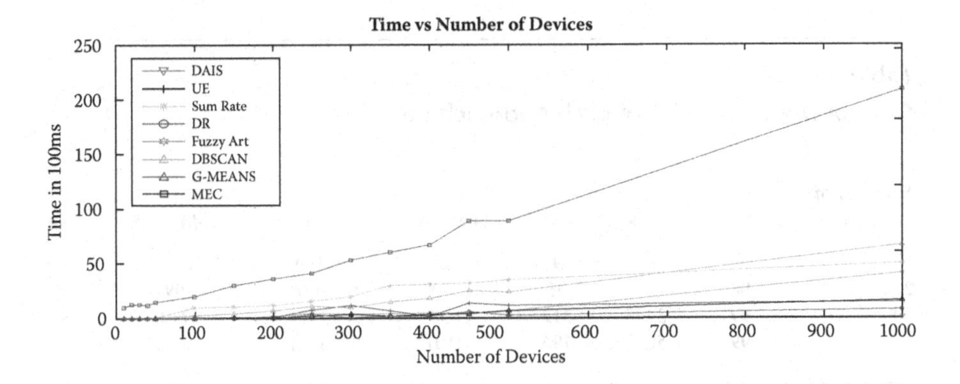

Figure 7.22 Time vs number of devices of different approaches.

Continuing with the DSR approach, as expected, it provides the best results, for both SE and PC, irrespective of the number of UEs used. This confirms that the enhancements made on DSR (see Section 6.2) improved it and made it to out-perform DAIS. However, due to excessive signalling needed, the execution time of DSR is slow, but still quicker than DBSCAN and MEC.

For the Fuzzy ART approach, in terms of SE, we see significant gains when 10–50 UEs are used and good gains for less than 500 UEs. For more than 500 UEs and less than 1000 UEs, Fuzzy ART ranks third. Also, in terms of PC, Fuzzy Art provides good performance when 50–500 UEs are used and even better with less than 50 UEs. Additionally, Fuzzy Art from 500 to 1000 UEs achieves medium range performance and it takes the third and fourth place accordingly. In terms of total execution time, Fuzzy ART ranks fourth.

For the DBSCAN approach, its performance is somewhere midway of all others in terms of SE and PC. In terms of total execution time, DBSCAN ranks second-last.

For the G-MEANS approach, we can see significant gains from 10 to 50 UEs and also for more than 500 UEs, but its performance is not consistent in the whole range of 10–1000 UEs. In terms of total execution time G-MEANS ranks third.

For the MEC approach, for less than 20 it provides the best results in terms of SE and PC, also achieving good results until 200 UEs. Above 200 UEs it offers decreasing performance. In terms of total execution time MEC ranks last.

In summary, based on the aforesaid, the approaches that provide the best results in terms of SE are DSR (1st), G-MEANS (2nd), and DAIS (3rd). The worst results are provided by DR. On the other hand, the approaches that provide the best results in terms of PC are DSR (1st), DAIS (2nd), and G-MEANS (3rd). The worst results are provided by DR. In terms of time execution time, the approaches that provide the best results are DAIS (1st), DR (2nd), and G-MEANS (3rd). The worst results are provided by MEC.

Overall, all approaches show a significant variation in SE and PC performance as the number of UEs change. This can be observed in the following statistics, over the range of 10–1000 devices:

Table 7.11
Control Decision Delay of each Approach (ms)

Number of Devices	DAIS	DSR	DR	FuzzyART	DBSCAN	G-MEANS	MEC
50	**9**	99	**9**	**18**	**100**	**9**	1512
200	**98**	1185	**95**	**223**	697	**99**	3620
500	**99**	3495	**312**	**698**	2412	**712**	8912
1000	**99**	5012	**796**	4101	6734	**1634**	20,905

1. The minimum percentage change of SE is: (i) 36.34% for DSR; (ii) 28.35% for G-MEANS; (iii) 24.87% for DAIS (Eq. 7.20 for 1000 UEs).
2. The maximum percentage change of SE is: (i) 36.34% for DR; (ii) 33.00% for DBSCAN; (iii) 30.90% for MEC (Eq. 7.9 for 1000 UEs).
3. The minimum percentage change of PC is: (i) 68.87% for non-D2D UE; (ii) 49.60% for DBSCAN; and (iii) 49.50% for DR (Eq. 7.11 for 1000 UEs).
4. The maximum percentage change of PC is: (i) 68.87% for DSR; (ii) 56.35% for G-MEANS; and (iii) 53.26% for Fuzzy ART (Eq. 7.21 for 1000 UEs).

Overall in terms of execution time, the faster approach is DAIS (DAI) irrespective of the number of UEs used (from 10 to 1000 UEs). The slowest approaches are MEC, DBSCAN (centralized) and DSR (distributed). The execution time observations are shown in Figure 7.22 and Table 7.11.

7.1.7 CONCLUDING REMARKS ON PERFORMANCE EVALUATIONS

At our first evaluation with the Sum Rate and the initial instance (non-enhanced) of DAIS, we show that the initial instance (non-Enhanced) DAIS achieves the same SE with sum rate and DR with non-D2D-UE however with less time. Additionally, we show that is consumes less PC than all other approaches.

Continuing at our second evaluation. The performance evaluation focused on the efficiency of SE and PC and their tradeoff regarding the TP, while respecting QoS and QoE. In all investigated approaches, the results showed that by reducing the TP of communication the SE and PC of the network in less than 100 UEs, is acutely affected. In contrast with more than 100 UEs, the SE is not highly affected, but the PC is always drastically affected in the sense of a reduction.

Furthermore, we compared the efficiency of each approach[11] in terms of SE and PC, cluster formation, signalling overhead (i.e., volume of messages exchanged) and control decision delay (see Table 7.12). Our findings show that the enhanced DSR outperforms, in terms of SE and PC, all other approaches. Then, in terms of SE, G-MEANS and DAIS outperform the other approaches and in terms of PC the Fuzzy ART, G-MEANS and DAIS outperform the other approaches. In terms of clusters

Table 7.12
Efficiency of each Approach in Terms of SE, PC, Clustering, Control Decision Delay and Signalling Overhead

	SE	PC Efficiency	Clustering Efficiency	Decision Delay	Signalling Overhead Efficiency
DAIS	Very good	Very good	Excellent	Excellent	Excellent
non-D2D UE	Good	Poor	N/A	N/A	N/A
DSR	Excellent	Excellent	Good	Average	Poor
DR	Average	Average	Average	Very good	N/A
FuzzyART	Very good	Very good	Good	Good	Average
DBCAN	Average	Average	Poor	Poor	Very good
GMEANS	Very good	Very good	Average	Very good	Good
MEC	Good	Good	Average	Poor	Very good

and messages needed for each approach to finish transmission mode selection, all approaches create clusters in the most 'accurate' positions with the use of WDR (in DAIS), Sum Rate (in DSR) and Data Rate (in Fuzzy ART, MEC, DBSCAN, G-MEANS) measurements. The results showed that DAIS achieves the most accurate clusters in the least time (see also Table 7.8). More specifically, for the running instance when 1000 UEs are used, DAIS is the fastest with a total execution time of around 100 ms, followed by DR with a total execution time of around 800 ms and by G-MEANS with a total execution time of 1600 ms. The slowest approaches are MEC, DBSCAN (centralized) and DSR (distributed). In our opinion, for a deployable D2D implementation, time is one of the most important evaluation metric along with SE and PC.

Additionally, the D2D Effectiveness, Stability, Productivity, and QoE and QoS fairness metrics were also investigated (see Table 7.13). The DSR approach is the only D2D effective (both in SE and PC) for all running instances (i.e., with 50, 200, 500, and 1000 UEs), whereas the approaches Fuzzy ART, DAIS, and MEC are D2D SE effective. Likewise, DSR is the only D2D Stable approach, whereas DAIS, Fuzzy ART and MEC are only SE stable. Moreover, DSR and DAIS are the only D2D Productive approaches. With regard to QoS Fairness metric, DAIS, non-D2D UE, DSR, DR and Fuzzy ART can be characterized as QoS fair. Also, regarding QoE fairness metric, all approaches are considered as fair[12] in terms of network resources usage (i.e., data rate).

Overall, our findings show that it is beneficial to use AI/ML approaches for transmission mode selection in 5G D2D communication by achieving energy conservation 5G requirement and mMTC, eMBB 5G use cases. The investigated approaches are fair and in some cases D2D efficient, stable, and productive (i.e., DSR, DAIS, Fuzzy ART). In terms of time of execution, the DAIS is the fastest approach and

Table 7.13
Characteristics of Each Approach in Terms of Fairness, Effectiveness, Stability and Productivity

	QoS Fair	QoE Fair	D2D Effective SE	D2D Stable PC	D2D Productive SE	PC	SE	PC
DAIS	✓	✓	✓		✓		✓	✓
non-D2D UE	✓	✓						
DSR	✓	✓	✓	✓	✓	✓	✓	✓
DR	✓	✓						
FuzzyART	✓	✓	✓		✓			
DBCAN		✓						
GMEANS		✓						
MEC		✓	✓					

DSR is the slowest. So, given these tradeoffs, the applicability of each approach must be determined by the evaluated use case requirements (e.g., a DSR implementation may be adopted in a stadium where there is a limited movement).

7.2 PERFORMANCE EVALUATION IN A DYNAMIC ENVIRONMENT

In this section, we consider a dynamic environment. Next we provide a description of: (i) the evaluation scenarios; (ii) the assumptions and terms used in the evaluation scenarios; (iii) the formulation of calculation of Spectral Efficiency (SE) and Power Consumption (PC) using Shannon Equation considering speed; iv) the problem description and formulation in a dynamic environment; (v) the methodology used for the performance evaluation; (vi) the simulation environment and its simulation parameters. Finally, it examines, evaluates, and compares the performance of DAIS and DSR with the Distributed Single Hop Relay Approach (SHRA) approach, considering dynamic network conditions (i.e., incorporating mobility, speed, direction, etc.) causing changes in the D2D network topology through subsequent Time Steps (TS) of execution. The difficulty there is that in each Time Step of execution the new selected transmission mode can affect existing clusters, as well as the formation of new clusters and backhauling links, that could result in disconnected/disjointed clusters. However, these clusters and paths should not be affected, even if the UE moves away from the Cluster Head (CH).

Thus, it evaluates how the SE and PC are affected in a dynamic environment, also against other competing approaches, such as Distributed Random, Distributed DSR, centralized non-D2D-UE and Distributed Single Hop Relay Approach (SHRA). The results obtained demonstrate the superior performance of DAIS over the SHRA, DSR, Distributed Random and non-D2D UE approach in terms of SE and PC. Also, it is shown that the expected signalling overhead and control delay in responding to changes of the dynamic network affects negatively the network performance (i.e., a decrease of the SE and increase of PC). Finally, it provides concluding remarks on the dynamic case.

7.2.1 PROBLEM DESCRIPTION, FORMULATION AND INVESTIGATED ASSOCIATED APPROACHES TO THE OPTIMIZATION OBJECTIVE

Our primary goal is to tackle the D2D challenges mentioned in Ref. [41] and Section 2.2.2, aiming the implementation of 5G/6G D2D communication in a dynamic environment. More specifically, our objective is to utilize our findings on the DAIS and DAI Framework BDIx agent to select the most appropriate transmission mode (i.e., D2DSHR, D2DMHR, D2D Client) to form a good backhauling network and good formation of clusters. By selecting the most appropriate transmission mode of a D2D Device, we seek to maximize the total SE jointly while minimizing the total PC through clustering and backhauling.

7.2.1.1 Assumptions and Terms

Our investigation considers the following assumptions:

- A single Base Station (BS) with a total number of N moving UEs (D2D Devices) forming the D2D communication network.
- A D2D network with a total number of Z devices representing the devices that share their link (i.e., D2DSHR, D2DMHR, BS).
- A D2D network that includes a total number of X devices representing the devices that are utilising the shared link and are attached as clients to Devices that share their link (i.e., D2D Client to D2DSHR, D2DSHR or D2DMHR to D2DMHR, D2DSHR or D2DMHR to BS). Please note that X includes the number of devices that connect to the BS.
- Each D2D device has calculated the Weighted Bandwidth. The Weighted Bandwidth ($WBW_{D2D^{\alpha}_{TMS}}$) of a D2D Device is the percentage bandwidth that a UE is using over the Base Station links.
- A connection scenario with a single-antenna and a point-to-point communication.
- A Free Space Path Loss model (for calculating average received power)
- A basic noise model, the Additive White Gaussian Noise (AWGN), for calculating the signal-to-noise ratio and then the signal to interference plus noise ratio.
- An uplink scenario
- A scenario that D2DSHR shares over Wi-Fi and D2DMHR over LTE Direct Mobile Frequencies in an overlay fashion.
- A well defined D2D security protocol. The D2D security protocol is necessary for the D2D Devices to access the D2D communication and Telecom network securely. Additionally, it is needed to access the LTE ProSe service and guarantees access to all the features provided by the operator.
- In each TS, all D2D client devices have a pre-specified speed and direction set randomly from the beginning.

- In each TS, each D2D Device with D2D client mode randomly selects a speed among speed threshold and the pre-specified speed assigned (from before) according to TS before the run of transmission mode selection. After the run of transmission mode selection, if the device is selected to be a D2D client, it resets its speed to the pre-specified speed of the TS. The reason is for our simulation to be more dynamic and to show the potentials of each approach. Another reason is to evaluate each approach in terms of SE and PC per speed in each TS.
- When the simulation is initiated (TS = 0, as shown in Section 7.2.2.1), all devices have speed below or equal to the speed threshold (e.g., pedestrian speed). Also, the D2D Devices that selected transmission mode as D2D Relay or D2D Multi-Hop Relay at that time step, in the subsequent runs they do not change speed and transmission mode, while the rest of D2D Devices (that are D2D clients) can. Note that in any TS, when a D2D client selected a speed equal to the speed threshold (e.g., pedestrian speed) and by using the DAIS Plan (as shown in the Algorithm 6.2) changed the transmission mode to D2DSHR or D2DMHR, in the subsequent runs it can not change speed and transmission mode. The above assumption is used for all the investigated approaches in order to be fair in the evaluation.
- In this investigation we do not consider the Doppler effect [42].

Note:

- The number of D2D Clients connected to D2DSHR is restricted to 200 (Wi-Fi Direct).
- The number of D2D-Relay connected to D2DMHR is restricted to 1 (LTE Direct)
- The maximum Distance between D2DSHR and D2D Client is 200 m.
- The maximum Distance between D2DMHR, D2DSHR to D2DMHR is 600 m.

7.2.1.2 Spectral Efficiency and Power Consumption

In this section, we show the optimization problem in terms of the maximization of SE (Eq. 7.36) having as result the minimization of PC (Eq. 7.37) in a dynamic environment, considering the above assumptions, and Table 7.14.

The SE is derived from the Shannon–Hartley theorem (Eq.7.22) in (bits/s/Hz).

$$\mathrm{SE}_{\mathrm{link}}(\mathrm{D2D}) = \frac{C}{B} = \log_2\left(1 + \frac{S}{N}\right) \tag{7.22}$$

Given the Additive White Gaussian Noise (AWGN) as a basic noise model, considering a power- and bandwidth-limited scheme, and a Free Space Path Loss model, we calculate the SE from the channel capacity in Eq. 7.23.

$$\begin{aligned} \mathrm{SE} &= \frac{C_{\mathrm{AWGN}}}{W} = \log_2(1 + \mathrm{SNR}) \\ \text{where } \mathrm{SNR} &= \frac{\bar{P}}{N_0 W} \end{aligned} \tag{7.23}$$

Table 7.14
Parameters Description

Parameter	Parameters Description
C	Capacity (in bits per second b/s)
B	Bandwidth (in Hertz Hz)
S	Signal power (in mini Watts mW)
N	Noise power (in decibel dB)
C_{AWGN}	Capacity with the use of the Additive White Gaussian Noise (AWGN) noise model
W	Bandwidth (in bits per second bps)
SNR	Received signal-to-noise ratio (SNR)
N_0	Noise (in Watts per Herz W/Hz)
$\bar{P}$	Average received power (in mini Waatts mW) Calculated using a Free space model and a free space path loss
TP	Transmission power known to the channel (from the UE and Base Station specifications)

The PC in mW is given in Eqs. 7.24 and 7.25.

$$PC = TP - \bar{P} \tag{7.24}$$

$$\bar{P} = \frac{TP}{10^{\tau/10}} \text{ where } \tau \text{ is the path loss} \tag{7.25}$$

7.2.1.3 Problem Formulation

In our approach, the mobile system is considered as an uplink D2D Orthogonal Frequency-Division Multiple Access (OFDMA) cellular network that consists of the deployment of D2D Relays that act as Cluster Heads, D2D Multi-Hop Relays that act as intermediate nodes in backhauling links, and D2D Client Devices that connect to D2D Relay Devices in a dynamic environment where the devices have speed and direction. Therefore, in the network architecture, each D2D Relay serves as CH and shares its bandwidth with the use of Wi-Fi Direct. Additionally, the D2D Multi-Hop Relays serve as intermediate nodes of the backhauling towards the gateway (i.e., BS) that provide better bandwidth and connection links; the protocol that the D2D devices use in backhauling links is the LTE Direct. The direct connections towards BS are regular mobile connections, so LTE Direct is not required.

The network environment is considered to be an enterprise or domestic, that comprises N D2D Devices. There are three cases of link sharing type with X total number of user clients in the system. The cases are: (i) when the transmission mode selection is D2DSHR and acts as CH, it serves a maximum number of 200 users (Wi-Fi Direct restriction) and can accept connections from other devices of the D2D client (D2DC) mode; (ii) when transmission mode selection is D2DMHR, the maximum

Table 7.15
Terms Used in the Equations of Dynamic Problem Formulation

Term	Explanation
$\mathrm{D2D_{TMS}}$	(TMS for Transmission Mode Selection) All the devices that shares a link (i.e., select D2DSHR,D2DMHR Transmission Mode and BS)
o	$o \in 1,2,\ldots,N$, N is the total number of N moving UEs
ι	$\iota \in 1,2,\ldots,Z$, Z is the number representing the devices that share their link
κ	$\kappa \in 1,2,\ldots,X$, X total number of devices representing the devices that are utilising the shared link and are attached as clients to devices that share their link
$\mathrm{D2D^{\iota}_{TMS}}$	The examined device that shares a link (i.e., select D2DSHR,D2DMHR Transmission Mode and BS)
$\mathrm{D2D^{\iota}_{TMS}}C$	The number of all the client devices that connect to a specific device that share a link (i.e.,D2D Client to D2DSHR,D2DSHR or D2DMHR to D2DMHR, D2DSHR or D2DMHR to BS)
β	$\beta \in 1,2,\ldots,\mathrm{D2D^{\iota}_{TMS}}C$, $\mathrm{D2D^{\iota}_{TMS}}C$ is shown above
$\mathrm{D2D^{\iota}_{TMS}}(\beta)(o)$	The client devices that connect to a specific device that share a link (i.e.,D2D Client to D2DSHR, D2DSHR or D2DMHR to D2DMHR, D2DSHR or D2DMHR to BS)
$\mathrm{D2D^{\iota}_{TMS}}S$	The speed of the device that shares a link
$\mathrm{D2D^{\iota}_{TMS}}(\beta)(o)^D$	The distance of the device ($\mathrm{D2D^{\iota}_{TMS}}(\beta)(o)$) from the device that shares a link ($\mathrm{D2D^{\iota}_{TMS}}$)
$\mathrm{D2D^{S}_{TMS}}(\max)$	The maximum speed that the device that shares a link can have
$\mathrm{D2D^{D}_{TMS}}(\max)$	The maximum speed that the device that shares a link can have. The device have another link that connects to another device that shares its link towards BS
$\mathrm{BW_{link_{BS}}}$	The bandwidth of a UE (that is not D2D) link towards BS
$\mathrm{WBW_{D2D^{\iota}_{TMS}}}$	It is the % of the $\mathrm{BW_{link_{BS}}}$ bandwidth. The purpose is to have a ratio for comparison among the Data Rate of the D2D Devices

number of clients that the device can share is one (LTE Direct restriction) and can accept connection from another device of mode D2DMHR or D2DSHR; and (iii) when the shared device is the BS, it can serve more than one and less than N devices of mode D2DSHR and D2DMHR devices, or serve every other UE under the mobile network.

Thus (see Table 7.15 for the explanation of the terms used), $\mathrm{D2D^{\iota}_{TMS}}(\beta)(o)$, $\beta \in \mathrm{D2D^{\iota}_{TMS}}C$ *in* X and $\iota \in Z$ represents the user/client attached to $\mathrm{D2D^{\iota}_{TMS}}$ sharing a device (i.e., D2D client attached to D2DSHR, D2DSHR attached to D2DMHR, D2DSHR attached to BS, D2DMHR attached to other D2DMHR, D2DMHR attached to BS). Note that "TMS" in the $\mathrm{D2D^{\iota}_{TMS}}(\beta)(o) \mid \mathrm{D2D^{\iota}_{TMS}}$ represents the selected mode of the device and it can take the values D2DSHR, or D2DMHR or

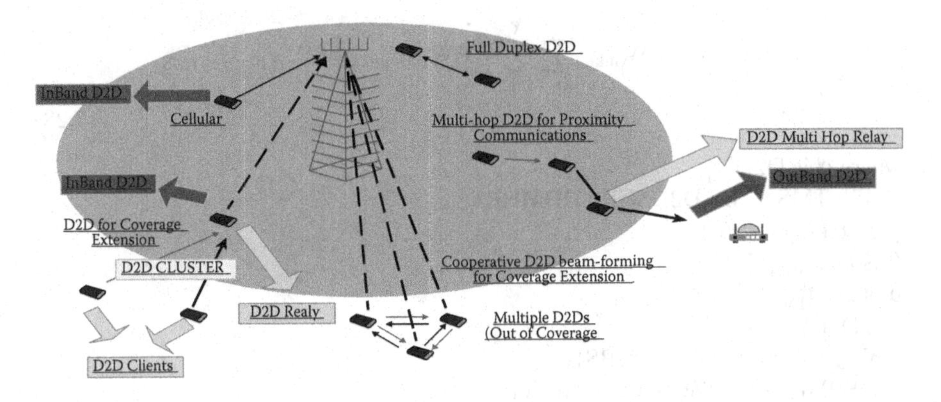

Figure 7.23 The D2D-Relays Are the Local Entries

BS. Also,

$\text{D2D}^{\iota}_{\text{TMS}} = \text{D2D}^{\iota}_{\text{TMS}} \mid \forall\, \text{D2D}^{\iota}_{\text{TMS}}$ and $\iota \in 1,2,\ldots,Z,$
$\text{D2D}^{\iota}_{\text{TMS}}(\beta)(o) =$
$\text{D2D}^{\iota}_{\text{TMS}}(\beta)(o) \mid\forall\, \text{D2D}^{\iota}_{\text{TMS}}(\beta)(o)$ where
$\beta \mid \forall\, \beta \in 1,2,\ldots,\text{D2D}^{\iota}_{\text{TMS}}C$
$\iota \mid \forall\, \iota\ \in 1,2,\ldots,Z$ and $o \mid \forall\, o\ \in 1,2,\ldots,N$
The network system described above also includes a local entity (shown as D2D-Relay in Figure 7.23) that acts as the control unit that resolves the conflicts (in terms of interference) among D2D Relays client devices (D2D Client) with the use of the Wi-Fi Direct protocol. Additionally, the LTE Direct frequencies are assumed to use orthogonal resources to the macro-BS with the use of the preassigned by the BS frequency band; thus, the problem of Intercarrier interference (ICI) between the D2D Relays, D2D Multi-Hop Relays, and the macro-BS is not addressed but handled by the connection protocols. This is consistent with the self-autonomy envisioned for D2D Devices. In the D2D communication network, the problem of network optimization with the use of the correct transmission mode selection can be translated to a weighted sum rate maximization problem where the purpose is to increase the sum rate while keeping the PC of the network to a minimum.

In order to tackle the problem, we convert the weighted sum rate maximization problem to a SE maximization problem. So, our objective is to maximize the SE (i.e., Total Sum Rate;[13] see Eqs. 7.28 and 7.36) while keeping the PC (see Eqs. 7.37 and 7.29) to a minimum, through the transmission mode selection. The data rate of a link is estimated using Eqs. 7.26 and 7.27.

$$\text{Data Rate}_{\text{link}}(\text{D2D}) = \text{BW}_{\text{Link}}\frac{C}{B} = \text{BW}_{\text{Link}}\log_2\left(1+\frac{S}{N}\right) \tag{7.26}$$

$$\text{Data Rate}_{\text{link}}(D2D) = \text{BW}_{\text{Link}} \cdot \text{SE}_{\text{link}} \tag{7.27}$$

$$\text{Total}_{\text{SR}} = \text{BW}_{\text{link}_{\text{BS}}} \sum_{\iota=1}^{Z} \sum_{\kappa=1}^{X} \text{WBW}_{\iota} \text{SE}_{\text{link}}(\beta) \text{BV}_{\alpha,\beta,\gamma,\delta,\varepsilon,\text{TMS}} \tag{7.28}$$

where α is $\text{D2D}^{\iota}_{\text{TMS}}$
$\text{TMS}|\forall\text{TMS} \in$ BS,D2DSHR,D2DMHR
γ is $\text{D2D}^{\iota}_{\text{TMS}}(\kappa)(o)$
β is $\text{D2D}^{\iota}_{\text{TMS}}C$
δ is $\text{D2D}^{\iota}_{\text{TMS}}S$
ε is $\text{D2D}^{\iota}_{\text{TMS}}(\kappa)(o)^{D}$
BW_{link} is Link Bandwidth (BS)
and WBW_{ι} is the Weighted Bandwidth
of D2D^{ι}_{TMS} in conjunction with the BW_{link}

$$\text{Total PC} = \sum_{j=1}^{N} \text{PC} \tag{7.29}$$

Overall, the optimization problem is to find the optimal Transmission Mode, considering a dynamic environment, in order to maximize the total sum rate with the selection of the best transmission mode that has as a result the minimization of Total PC, as follows:

$$\text{BV}_{\text{D2D}^{\iota}_{\text{TMS}}(\kappa)(o),\text{D2D}^{\iota}_{\text{TMS}}C,\text{D2D}^{\iota}_{\text{TMS}}S,\text{D2D}^{\iota}_{\text{TMS}}(\kappa)(o)^{D},\text{TMS}} \in 0,1,\forall\ \text{D2D}^{\iota}_{\text{TMS}} \tag{7.30}$$

$$\text{D2D}^{\iota}_{\text{TMS}}(\kappa)(o)\ \kappa \in X, o \in N, \text{D2D}^{\iota}_{\text{TMS}} \in Z \wedge \tag{7.31}$$

$$\text{D2D}^{\iota}_{\text{TMS}}C \in 1,\ldots 200, N \wedge \tag{7.32}$$

$$\text{D2D}^{\iota}_{\text{TMS}}S \leq \text{D2D}^{S}_{\text{TMS}}(\max) m/s \wedge \tag{7.33}$$

$$\text{D2D}^{\iota}_{\text{TMS}}(\kappa)(o)^{D} \leq \text{D2D}^{D}_{\text{TMS}}(\max) \tag{7.34}$$

where $\text{SE}_{\text{link}}(\iota)$ is directly related to SNR (Eq. 7.23)
and

$$\sum_{n=1}^{N} \bar{P}(n) \leq P_{\text{D2D}^{\iota}_{\text{TMS}}}(\max) \tag{7.35}$$

$$\text{Max Total SR} = \max \text{Total}_{\text{SR}} \tag{7.36}$$

$$\text{Min Total PC} = \min \sum_{j=1}^{N} \text{PC} \tag{7.37}$$

The binary variable (BV) of Eq. 7.30 corresponds to the transmission mode selection of the D2D Device and the allocation decision of the Device to another D2D Device that share its link (e.g., D2DSHR, D2DMHR or BS) where

$$o|\ \forall\, o \in 1,2,\ldots,N,\ \kappa||\ \forall\kappa \text{ in } 1,2,\ldots,X \text{ and } \iota|\ \forall\, \iota\ in\ 1,2,\ldots,Z.$$

More specifically, when the device selects to be a D2D client ($D2D^{\iota}_{TMS}(\kappa)(o)$) to a specific link sharing device ($D2D^{\iota}_{TMS}$), some constraints must be satisfied in order for the BV to result in "1", targeting towards maximization of the Sum Rate. In terms of constraints: (i) the number or already connected devices to the sharing device are subject to the constraint given by Eq. 7.32; (ii) the speed of the sharing device is subject to constraint given by Eq. 7.33; (iii) the distance among the sharing and the D2D client device is subject to the constraint given by Eq. 7.34; and (iv) the presence of inter-cell interference subject to the power constraint given by Eq. 7.34.

Furthermore, when the device is a D2D-Relay connected to D2DMHR, forming a backhauling link, the D2DMHR device is subject to the constraint given by Eq. 7.32 on how many devices they can connect. Basically, based on this constrain, only one D2D-Relay can connect to and associate with the D2DMHR. So, for the rest of devices that try to achieve connection to the D2DMHR, the BV will return 0. Moreover, each client D2D Device's channel PC $\bar{P}$ is considered for the D2D sharing Device ($\in$ D2DSHR,D2DMHR,BS) transmit power on the specified channel connection according to the limitation in Eq. 7.34.

Note that the data rate is considered weighted, according to our formulation, for two reasons: (i) due to different technologies that the device can use according to the transmission mode that is selected (e.g., Wi-Fi Direct to share over D2D Clients, LTE Direct to share a link to other D2D-Relays); and (ii) because the D2D-Relay device shares a fraction of its link bandwidth $WBW_{D2D_{TMS}}$ with its clients. This fraction of bandwidth is calculated as a percentage of the maximum achievable bandwidth in the network according to the protocol used (i.e, Wi-Fi Direct or LTE Direct).

Hence, the optimization problem is to maximize the weighted sum rate over the network in the presence of inter-cell interference subject to: (i) power constraint per node "o" as in Eq. 7.34 and intra-cell orthogonal allocation; (ii) number of client devices constraint according to Eq. 7.32; (iii) speed constraint according to Eq. 7.33; and (iv) distance constraint according to Eq. 7.34. Overall, the generic weighted sum rate maximization problem as described in Eq. 7.36 and Eq. 7.28 is a non-convex optimization problem with nonlinear constraints shown to be NP-hard (see Ref. [43]).

In the next section, to solve the problem, we implement in a heuristic way a specific DAI framework and Plan considering a dynamic environment and thereafter evaluate its performance. To further simplify the problem, in our approach, we examine the SE by setting the WBW_{ι} to "1". As a result, we accept that the Weighted Bandwidth rate[14] among the Wi-Fi Direct, BS Link and LTE Direct is the same. Therefore, our equation is simplified as in Eqs. 7.38 and 7.37. Additionally, the assumptions mentioned above and constraints on the calculation of SE are considered in our system. So, the optimum sought Total SE (Eq. 7.38) that will have as a result a decrease of the Total PC (Eq. 7.37) is given by:

$$\text{Max}^{\cdot}\text{Total}^{\cdot}\text{SE} = \max \sum_{\iota=1}^{Z} \sum_{\kappa=1}^{X} SE_{\text{link}}(\beta) BV_{\alpha,\beta,\gamma,\delta,\varepsilon,\text{TMS}} \tag{7.38}$$

7.2.2 PERFORMANCE EVALUATION

This section examines, evaluates, and compares the efficiency of DAIS, DSR, SHRA, and non-D2D UE under a D2D communications network with a range (10...1000) number of UEs in a dynamic D2D communication network setting. In addition, this examination considers the random change of speed and direction, hence proximity among the D2D Devices.

Table 7.16 shows the type of control performed and network knowledge needed by each approach mentioned above (DAIS, DSR, Distributed Random and non-D2D UE) along with the elaborated SHRA.

7.2.2.1 Methodology

Our examination focuses on the dynamicity of the mobile network. Consequently, we consider changes in the Transmission Power (TP), speed and direction of the UEs, number of devices in the network, and changes in the D2D network topology through subsequent TS of execution. Our examination specifies a Time Step (TS) of 100 ms (this is empirically selected to give a fast response for the given speed dynamics). TS = 0 relates to the initial D2D network topology. TS = 1 relates to the network topology after 100 ms, TS = 2 to the network topology after 200 ms, and so on. We evaluate the investigated approaches with maximum execution of TS = 5 at 1000 UEs. Additionally, to be fair with the time of execution, all approaches, except the DSR due to its large execution time (as shown in Refs. [25, 26, 44] and Section 7.1.6.6), are executed every 100 ms (i.e., every TS) to adapt to the transmission mode of the D2D devices based on the changes occurred on the D2D network topology.

To simplify the investigated problem, those D2D Devices that were initialized in TS = 0 to D2D-Relay mode will keep the same transmission mode (D2D-Relay) and speed (e.g., pedestrian speed) during all TSs of execution. Additionally, for the rest of the D2D Clients, if they decide to become D2D-Relays in the subsequent TS they need to keep the same transmission mode (D2D-Relay) and speed (e.g., pedestrian speed) during all TSs of execution. Also, for the DAIS approach, we assume that the

Table 7.16
Evaluated Approaches Type of Control & Network Knowledge They Need

Approach(es) Investigated	Type of Control	Network Knowledge
DAIS	DAI (distributed, decentralized)	Local knowledge
Distributed random	Distributed	Global knowledge
SHRA	Distributed	Reduced knowledge
DSR	Distributed	Global knowledge
non-D2D UE	Centralised	Global knowledge

BDIx agents accept any suggestion/proposal from another agent and the suggested action from the other agent is aligned with the agent's Desires. So, the agent replies with an "accept" message in each proposal, and executes the required actions.

For the DSR, we have from previous examinations (Section 6.2.1 and [25,26,44]) specific delays in the time of executions that makes the approach inappropriate for dynamic environments. More specifically, with DSR, when the number of devices in the network increase, the execution time needed for deciding on the transmission mode selection is increased as well. This makes the DSR not fast enough to be ready for recalculation after a specific Time Step with the network topology changing rapidly, resulting in degradation of SE and PC. The table of the different TS execution according to the number of devices in the D2D network is shown in Table 7.17. According to this table, the DSR runs for the first time with the initial D2D network topology at TS = 0 (initial step) for all UEs. Then, it runs a second time at TS = 1 to accommodate any changes on the network topology for a device range of 10–50. Afterwards, it takes more than the upper limit of our investigation of 5 TS to finish execution and conclude (as shown in Table 7.17).

Also, the SHRA (Section 6.3), the Distributed Random clustering approach (Section 6.4), the non-D2D UE Approach (Section 6.4), the DSR (Section6.2.5) and the DAIS Plan/algorithm (Section 6.1.6) are compared in terms of SE and PC by taking under consideration the dynamics of the Mobile Network. These relate to changes in the transmission power, UE speed, UE direction, number of devices in a D2D communication network, and network topology in different TS of execution.

As a starting point (i.e., TS = 0), we set the initial values of UE speed to 15 m/s, transmission power to 160 mW and UE direction to 90°. Afterwards, we rerun our simulation to examine the behaviour of the different approaches in subsequent TS (from TS = 1 to TS = 5) by changing a random parameter (e.g., speed, direction, transmission power) generated by a randomizer and increasing the number of UEs in the D2D network from 10 to 1000 UEs. In most of the evaluations, we examine the D2D network topology at TS = 5 and 1000 UEs cases. Also, the speed and direction are set at a constant 15 m/s and 90°, respectively.

Overall, in our investigation (as shown in Section 7.2.2.3.1), the following have been examined and demonstrated:

- The effect that the transmission power has on the dynamic DAIS, in terms of overall PC and total SE achieved over time with a variable number of Devices.

Table 7.17
DSR Time of Execution

DSR - Time of Execution

TS	0	1	10	12	35	50
UEs	All UEs (from BS)	10–50	100	200	500	1000

For the communication power, a "brute force" investigation was executed with values from 160 to 60 mW using a decreasing step of 10 mW.

- The behaviour and performance of the investigated approaches in terms of SE and PC considering the dynamics of the Mobile Network. These relate to changes in the Transmission Power, D2D network topology in different TS of execution, UE speed, number of devices in the network, and UE direction.

This investigation aims to examine and prove that: (i) the clusters created by all approaches and, more specifically, the dynamic DAIS plan algorithm using the WDR/DR as a metric are in the best positions; and (ii) the back-hauling links created with D2DMHR devices are helpful in a dynamic environment. Even changes in UE speed, UE direction, and D2D Network topology over the different TS of Execution do not heavily affect the resulting SE and PC.

7.2.2.2 Simulation Environment

In the simulation, a range of 10–1000 D2D Devices was used. The devices are placed in a cell range of 1000 m radius from the BS using a Poisson Point Process distribution model. We keep the same comparison measurements of performance (Total SE and Total PC), and the same equations/formulas for D2D UEs for battery power level estimation and WDR as in Chapter 6. However, the Total SE and Total PC of the D2D network is calculated as shown in Section 7.2.1, basically by adding all the achieved data rates of all nodes in the network.

For all approaches, the assumptions of the simulation are shown in Section 7.2.1.1. Also, the constraints are shown in Section 7.2.1.2 and the simulation parameters in Table 7.18. The DAIS and Sum Rate terms and parameters are shown in the Appendix.

The simulation environment is implemented in Java using specific libraries from MATLAB 2020a and more specifically the "5G/LTE Toolbox" [45] in conjunction with the JADE library (integrated with FIPA ACL and extended with BDI4JADE library) [46–51]. The hardware used for the simulation is the following: (i) an Intel(R) Core(TM) i7-8750H CPU @ 2.20 GHz; (ii) 24 GB DDR4; (iii) 1TB SSD hard disk; and (iv) NVIDIA GeForce GTX 1050 Ti graphics card with 4GB DDRS5 memory.

7.2.2.3 Results

In this section, we examine the effect that the transmission power (TP) has on the DAIS regarding total PC and total SE (i.e., Total Sum Rate). Also, we analyze the behaviour of the investigated approaches in terms of SE and PC considering the dynamics of the Mobile Network. This relates to changes in TP, D2D network topology in different TS of execution, UE speed, UE direction, and Number of Devices in the network.

Table 7.18
Simulation Parameters

Simulation Parameters	Value
D2D power	130 mW or otherwise defined [27–29]
UE power	260 mW or otherwise defined [27–29]
Wi-Fi direct radius	200 m [30]
LTE direct radius	600 m [31]
BS range	1000 m [27–29]
Path loss exponent (Urban Area)	3.5
BS antenna gain	40 dB [27–29]
UE/D2D antenna gain	2 dB [27–29]
PERC data rate	20% (≤200) and 35% (>200) [25,26]
Device battery threshold	75% [26]
MAX speed to form Backhauling	15 m/s
No	0.0001
D	200 Users
N (no of UEs)	10–1000
Shadowing	Log-normal
Mobility	Dynamic scenario

7.2.2.3.1 DAIS TP Examination Results

The effect that the TP has on DAIS, in terms of total PC and total SE (sum rate) achieved, is illustrated in Figures 7.24 and 7.25. In the scenario used, the TP is reduced from 160 to 60 mW, the amount of UEs are increased from 10 to 1000, while the speed (15 m/s) and direction (90°) of the UEs are kept constant. The results relate to the D2D network topology changes occur from TS = 0 to TS = 5 and examine how DAIS approach can react to changes related to the link TP and number of devices in the D2D network. So, we examine the effect that the TP, number of devices, and network topology changes in time that DAIS has in a dynamic environment, regarding total PC (i.e., for power reservation and green energy) and total SE (i.e., eMBB) targeting 5G requirements.

As observed in Figure 7.25, for all TS, by reducing the TP of the communication and increasing the number of UEs (D2D Devices), gains are provided on the PC with a small trade-off on the SE. Also, the gains mentioned earlier vs trade-offs can be seen in more extended ranges in networks with large numbers of devices (500, 1000). More specifically, for any number of UEs in all TS, the maximum percentage change observed in terms of SE is 22% and in terms of PC is 70%.

Additionally, we can see from the figures that there are some noticeable unexpected increments in measurements in terms of SE when we change the TP, at specific values. [15] These unexpected increments follow the same pattern at specific TP levels during each time step. The increments drastically affect the SE in the small number of devices (≤200). In our opinion, the above increments are related to an in-

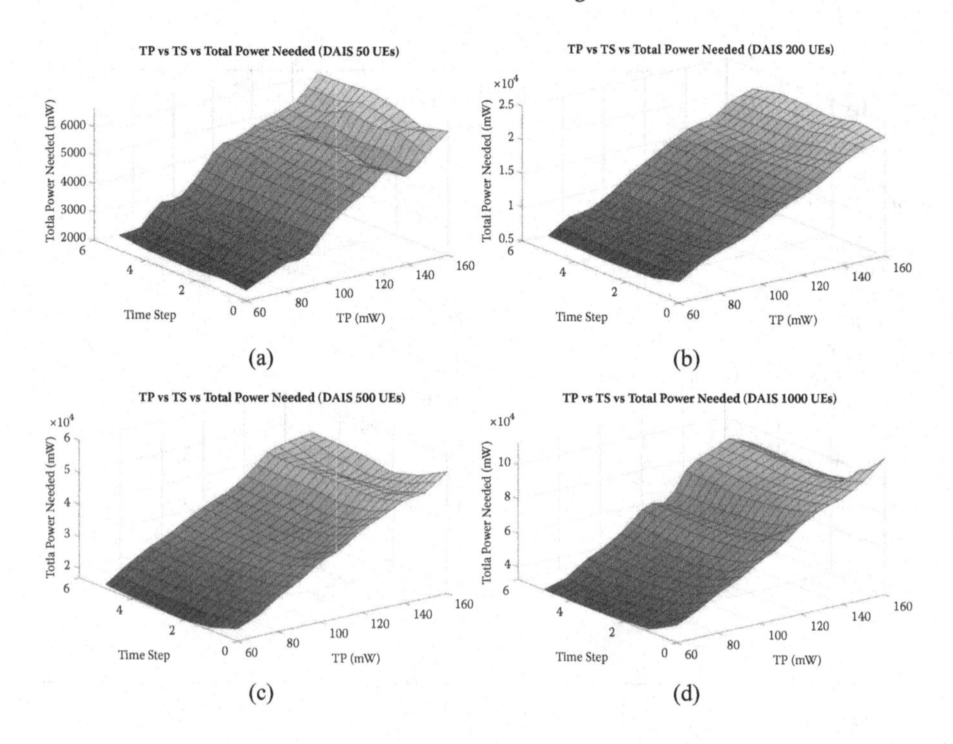

Figure 7.24 TP vs TS vs total PC: (a) 50 UEs, (b) 200 UEs, (c) 500 UEs, and (d) 1000 UEs.

crement of cluster numbers under the D2D network that, when reached, are restricted and reduced, along with the backhauling links, by the use of the WDR threshold (as shown in Section 6.1). More precisely, we have the following cases, per range of TP and number of UEs:

- From 90 to 100 mW TP with 50 UEs we have an increment of clusters from 7 to 19.
- From 90 to 100 mW TP with 200 UEs we have an increment of clusters from 49 to 106.
- From 130 to 140 mW TP with 50 UEs we have an increment of clusters from 6 to 9.
- From 130 to 140 mW TP with 200 UEs we have an increment of clusters from 59 to 160.
- From 110 to 120 mW TP with 500 UEs we have an increment of clusters from 99 to 201.
- From 110 to 120 mW TP with 1000 UEs we have an increment of clusters from 159 to 201.

Moreover, our examination showed that in terms of PC, the changes are smooth with no unsuspected increments. Another important observation is that DAIS appears unaffected in terms of SE and PC irrespective of any changes that occur on the TP, number of devices and TSs in a dynamic environment.

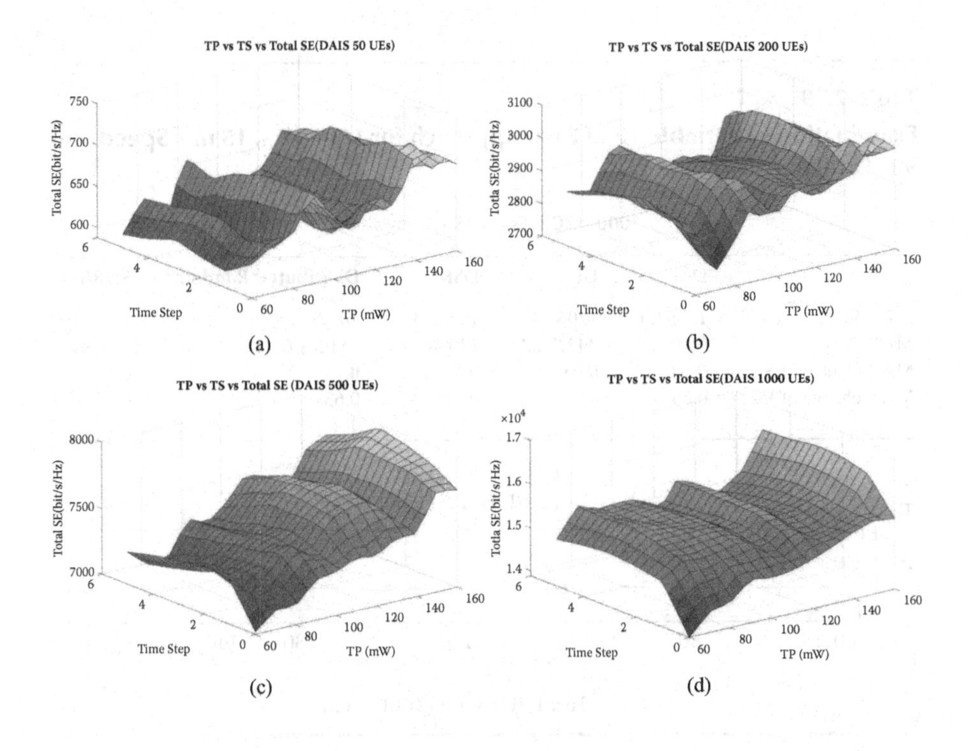

Figure 7.25 TP vs TS vs total SE: (a) 50 UEs, (b) 200 UEs, (c) 500 UEs, and (d) 1000 UEs.

7.2.2.3.2 Behaviour of the Investigated Approaches on Dynamic TP

This section examines the case where the TP is reduced from 160 mw to 60 mW, while the speed (15 m/s), the number of devices (1000 D2D Devices) and direction (90°) of the UEs are kept constant. The results relate to the D2D network topology at TS = 5 and examine how each approach can react to TP changes. Therefore, we evaluate the effect that the TP has in a dynamic environment at the investigated approaches regarding total PC (i.e., for power reservation and green energy) and total SE (i.e., eMBB) targeting 5G requirements.

As illustrated in Table 7.19 and Figure 7.26, in this investigation DAIS approach provides the best results in terms of SE and PC. Additionally, DAIS achieves the maximum PC reduction (followed by Sum Rate) and the minimum SE reduction (followed by the non-D2D UE approach) compared to all other related approaches. Please note that the number in bold represents the maximum value in the table while the values in italic represent the minimum value.

Table 7.19
Examination of Variable TP of Each Approach for 1000 UEs, 15 m/s Speed and 90° Direction

1000 D2D UEs - 5 TS - 15 m/s - 90°					
	DAIS	**UE**	**DSR**	**Distributed Random**	**SHRA**
MAX SE	**16354.4**	12062.4	13290.4	*10832.1*	11843.5
MAX PC	*96551.6*	**243778.3**	140987.0	151041.0	145399.1
MAX change of SE	*0.08*	*0.03*	0.11	**0.14**	0.11
MAX change of PC	**0.66**	*0.03*	0.63	0.63	0.63

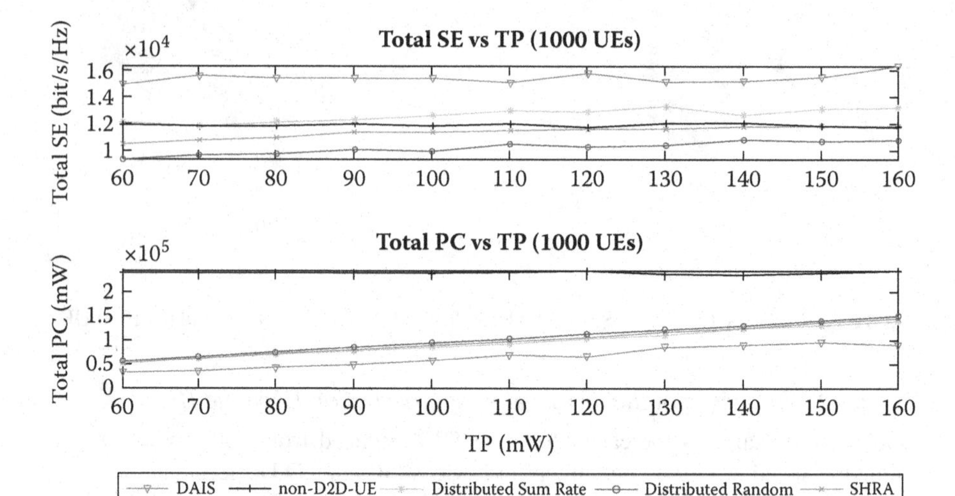

Figure 7.26 TP change investigation among the examined approaches.

7.2.2.3.3 Behaviour of the Investigated Approaches on Network Topology Changes Over the TS of Execution

This section examines the case where the TS is increased from 0 to 5 (which mainly relates to changes in D2D network topology), while the TP (160 mW), the speed (15 m/s), the number of devices (1000 D2D Devices) and direction (90°) of the UEs are kept constant. Therefore, we evaluate the effect that the Network Topology Change, via the TSs of Execution, has in a dynamic environment at the investigated approaches regarding total PC (i.e., for power reservation and green energy) and total SE (i.e., eMBB) targeting 5G requirements.

The performance of the investigated approaches is compared in terms of total SE (Sum Rate) and PC. The results are provided in Figure 7.27.

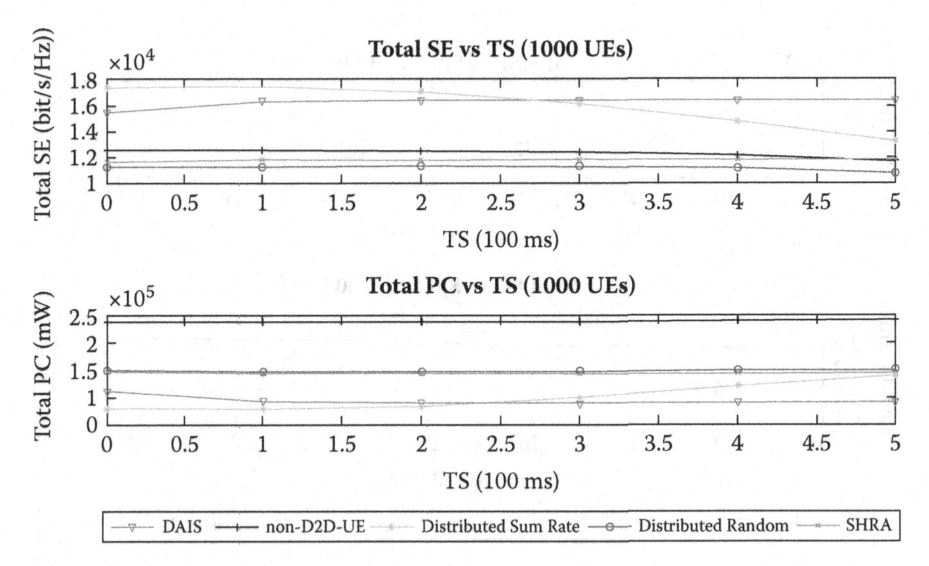

Figure 7.27 Total SE and total PC vs time steps with 1000 UEs, 15 m/s speed and 90° direction.

The best results from 0 TS until the 2.5 TS, in terms of SE and PC for 1000 devices, are provided by the DSR. These results have been achieved with the extension and the enhancements made, introducing the speed as an extension, Data Rate and Battery Power level thresholds as enhancements in the DSR to support dynamic networks. However, after 2.5 TS, the DSR degrades performance. The DSR does not keep the highest SE and PC values after 2.5 TS due to the large execution time (i.e., 50 TS) needed to decide on the transmission mode selection. This makes the DSR not fast enough to be ready for recalculation after 2.5 TS. For more details see Section 7.2.2.1, Table 7.17) and [25, 26]. The second-best performance, from 0 TS until the 2.5 TS, however very close to the one provided by Sum Rate, is achieved by DAIS. Non-D2D-UE, SHRA, and Distributed Random follow. After 2.5 TS, the best results in terms of SE are achieved by DAIS.

The results related to PC follows a similar pattern. The best results from 0 TS until the 2.5 TS are provided by the DSR, which, for the same reason described above, degrades performance after the 2.5 TS. After the 2.5 TS, the DAIS approach outperforms Sum Rate, followed by SHRA, Distributed Random and then non-D2D-UE.

Overall, what made DAIS outperform all other approaches in both SE and PC, are the adaptations and thresholds (i.e., speed, WDR, BPL) implemented (see Section 6.1.6), making DAIS capable to efficiently support dynamic environments (note that in our previous section work considering static environments as shown in Chapter 6 DSR and DAIS had the same SE).

Additionally, according to Figure 7.27, except for the DSR, all other approaches do not have any significant changes, in terms of SE and PC, over subsequent TS. More precisely, over subsequent TS, the DSR has a maximum SE reduction of 25% and a maximum PC increase of 45%.

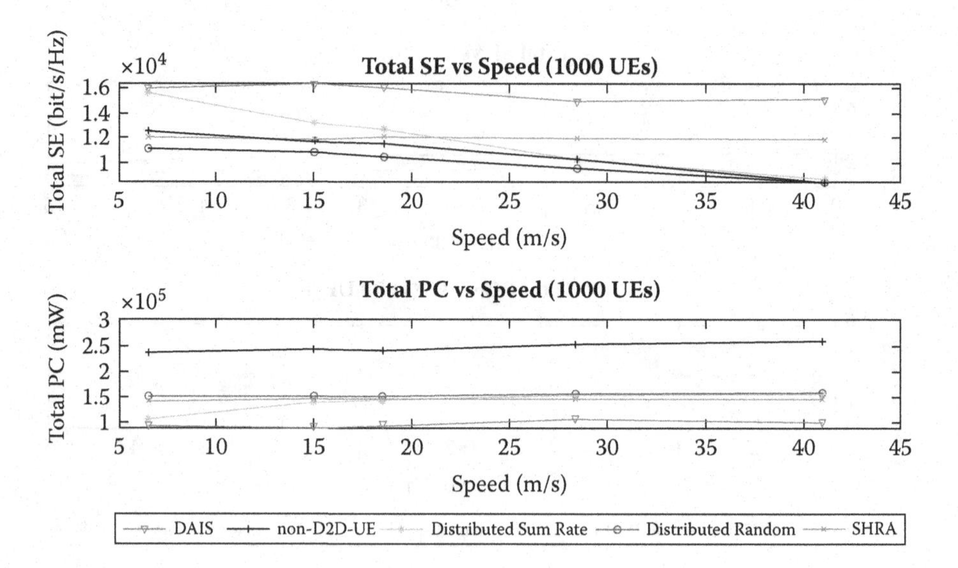

Figure 7.28 Total SE and total PC vs speed with 1000 UEs, at 5 TS and 90° direction.

7.2.2.3.4 Behaviour of the Investigated Approaches on Dynamic UE Speed

This section examines the case where the Speed of the UE changes randomly, while the TP (160 mW), the number of devices (1000 D2D Devices) and the direction (90°) of the UEs are kept constant. The results relate to the D2D network topology at TS = 5 and examine how each approach can react to the UE speed changes. The performance of the investigated approaches is compared in terms of total SE (Sum Rate) and PC. Consequently, we evaluate the effect that speed has in a dynamic environment at the investigated approaches regarding total PC (i.e., for power reservation and green energy) and total SE (i.e., eMBB) targeting 5G requirements. As shown in Figure 7.28), the best performance in terms of SE and PC is provided by DAIS followed by SHRA. Note that DAIS and SHRA, in contrast with Distributed Random, non-D2D UE and Sum Rate (that approach close to zero (0)), are the only two approaches that still provide good results in terms of SE as the speed of the UEs increases, justifying their ability to support dynamic mobile environments. Also, in terms of PC, only the DSR is highly affected by the UE speed.

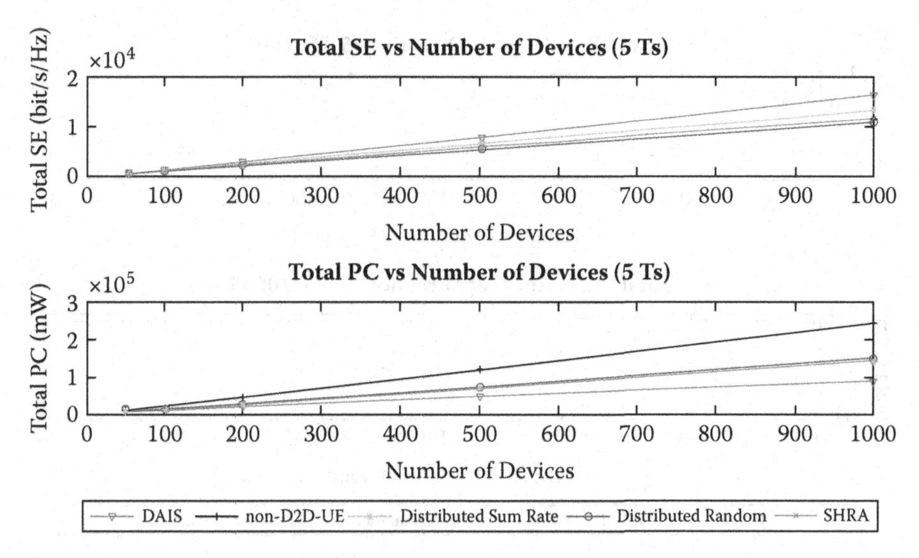

Figure 7.29 Total SE and total PC vs number of devices at 5 TS, 15 m/s speed and 90° direction.

7.2.2.3.5 Behaviour of the Investigated Approaches on Different Number of Devices in the Network

This section examines the case where the number of UEs in the network increases from 10 to 1000, while the TP (160 mW), the speed (15 m/s) and the direction (90°) of the UEs are kept constant. The results relate to the D2D network topology at TS = 5 and examine how each approach can react to the increasing number of UEs. Hence, we evaluate the effect that different number of devices have in a dynamic environment at the investigated approaches regarding total PC (i.e., for power reservation and green energy) and total SE (i.e., eMBB) targeting 5G requirements. As shown in Figure 7.29, the best performance in terms of SE and PC is provided by DAIS, irrespective of the number of devices in the network. The second-best performance in terms of SE is provided by the DSR, followed by the non-D2D UE, SHRA and Distributed Random. Additionally, the second-best performance in terms of PC is provided with the DSR, followed by the SHRA, the non-D2D UE and the Distributed Random approach.

7.2.2.3.6 Behaviour of the Investigated Approaches on Dynamic UE Direction

This section examines the case where the Direction of the UE changes randomly, while the TP (160 mW), the number of devices (1000 D2D Devices) and the speed (15 m/s) of the UEs are kept constant. The results relate to the D2D network topology at TS = 5 and examine how each approach can react to changes in the UE direction. So, we evaluate the effect that direction has in a dynamic environment at the investigated approaches regarding total PC (i.e., for power reservation and green energy)

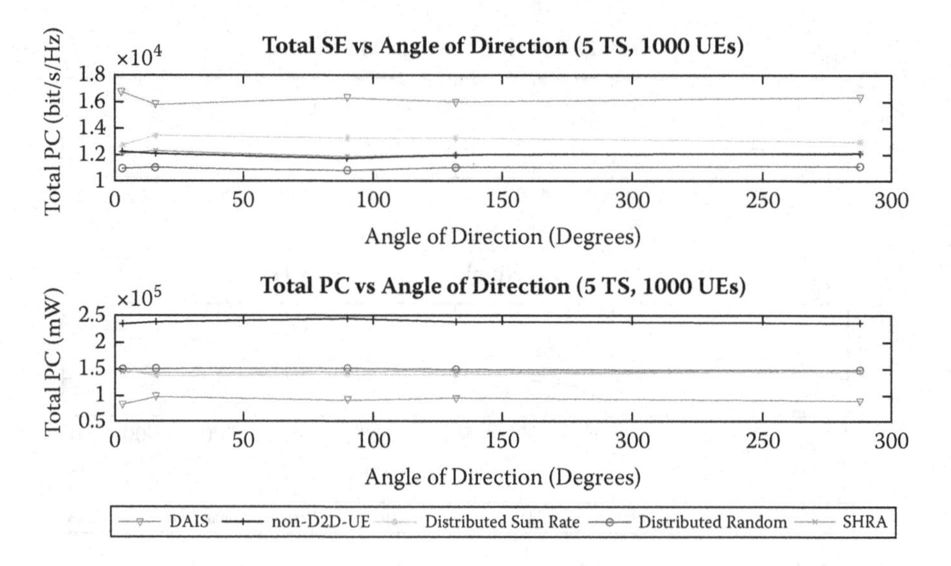

Figure 7.30 Total SE and total PC vs direction at 5 TS, 15 speed and 1000 UEs.

and total SE (i.e., eMBB) targeting 5G requirements. As shown in Figure 7.30), the best performance in terms of SE and PC is provided by DAIS, irrespective of the way the devices are moving in the network. The second-best performance in terms of SE is provided by the DSR, followed by the non-D2D UE, SHRA and Distributed Random. Additionally, the second-best performance in terms of PC is provided with the DSR, followed by the SHRA, the non-D2D UE and the Distributed Random approach.

7.2.2.3.7 Overall Observations

Overall, we examined the enforcement of the most significant thresholds such as the maximum speed to select a D2D-Relay, and the use of specific WDR (set to 20% when the number of UEs, ≤200) or 35% otherwise, and BPL, set to 75%, thresholds, as shown in Section 6.1.6). Additionally, as shown in Section 6.2.5, we enforce new thresholds for the DSR. These thresholds are related to the maximum speed to select D2D-Relay, the specific Data Rate that a D2D candidate device can connect to a D2D Relay (set empirically to 35%) and the Battery Power Level Threshold (that is set to 75%). Also, as shown in Section 6.3, in the case of the SHRA approach, we have made a slight change in the algorithm in order for the D2D Relay to receive multiple connections and not to be restricted by one (i.e., to allow the formation of clusters). The adjustments made on DAIS, Sum Rate and SHRA algorithms are implemented for achieving the maximum possible total sum rate (i.e., maximum SE) and maximum power reservation (i.e., minimum PC) in a range of 10–1000 of the number of devices in a dynamic environment.

We also analyzed the behaviour of the investigated approaches considering the dynamics of the Mobile Network. More specifically, we examined how each approach can react to the changes in UE speed and direction, causing variations in the D2D network topology, as well as to changes in the TP and number of Devices in the Network. Based on this examination, we compared the efficiency of each approach in terms of SE and PC. The results are summarized in Table 7.20.[16]

Overall, based on the results collected, the only approach that can provide excellent results in a dynamic environment, both in terms of SE and PC, is DAIS. More specifically, DAIS can react quickly to D2D Network topology changes caused through time (i.e., in the different TS), whether these are caused by variations in UE speed, UE direction, number of devices in the network or TP, and decide efficiently on the transmission mode that the D2D Devices will operate.

DSR comes second in terms of SE and PC. More specifically, in terms of SE, it provides "Good" results except for the case where network topology changes are caused due to variations in the UE Speed. In this case, the results provided are considered "Poor". Also, "Good" results are provided in terms of PC, except the cases where network topology changes are caused due to variations in the UE Speed and TP. In these cases, the results provided are considered "Average". Additionally, Sum Rate is the only approach that, in some cases, drops its SE and increases its PC drastically compared to all other approaches (see Figure 7.27). Thus, in our believe, if we introduce more TS in the simulation, more probably the DSR could conclude to be the last.

The SHRA approach, in terms of SE, in most cases is evaluated as "Average", except the case where network topology changes are caused by variations in the UE Speed. In this case, the results provided are considered "Good". Also, SHRA performance in TP variations is considered "Poor". Furthermore, SHRA performance in terms of PC is considered as "Average", except for the case where variations in UE Speed occur. In this case, the results of SHRA are "Good".

The Random approach, in terms of SE, provides "Poor" results in all respects. In terms of PC, the results provided are considered as Average except the cases where changes occur on the TP and the number of Devices in the D2D Network. In these cases, the performance of Random approach is "Poor".

Finally, the non-D2D UE approach, in terms of SE, provides "Average" performance, except in the case where changes occur on the UE speed. In this case, its performance is considered "Poor". In terms of PC, the performance of non-D2D approach is considered "Poor" in all respects.

7.2.3 CONCLUDING REMARKS ON DYNAMIC CASE

This section builds on our work presented in previous sections and it develops an extended version of DAIS, for selecting the D2D Transmission mode that the D2D Devices will operate in dynamic environments incorporating UE mobility and changes in the D2D Network topology. To set a benchmark and allow for a fairer comparison, we also extended and adapted: (i) the Distributed Sum Rate (DSR) approach, proposed in Chapter 6 and Section 6.2.4 to also support D2D Communication in

Table 7.20
Overall Evaluations of the Approaches Using the Dynamic Variables in Terms of SE and PC

Metric	SE					PC				
Investigation Approach	**DAIS**	**DSR**	**SHRA**	**Distributed Random**	**non-D2D UE**	**DAIS**	**DSR**	**SHRA**	**Distributed Random**	**non-D2D UE**
Transmission Power	Excellent	Good	Poor	Poor	Average	Excellent	Average	Average	Poor	Poor
Time step	Excellent	Good	Average	Poor	Average	Excellent	Good	Average	Average	Poor
Speed	Excellent	Poor	Good	Poor	Poor	Excellent	Average	Good	Average	Poor
Number of devices	Excellent	Good	Average	Poor	Average	Excellent	Good	Average	Poor	Poor
Direction	Excellent	Good	Average	Poor	Average	Excellent	Good	Average	Average	Poor

dynamic environments; and (ii) the SHRA approach [52], to additionally allow the D2D-Relays to accept more than one connections (i.e., create clusters). Furthermore, an extensive comparative evaluation of the enhanced DAIS, DSR, SHRA, Distributed Random and non-D2D UE is provided. During this evaluation, we analyzed the behaviour of the investigated approaches considering the dynamics of the Mobile Network and comparatively evaluated their performance, in terms of SE and PC, against a number of metrics. More specifically, we examined how each approach can react to the changes in UE speed and direction, causing variations in the D2D network topology, as well as to changes in the TP and number of Devices in the Network.

Overall, the results obtained demonstrated superior performance of DAIS over the SHRA, DSR, Distributed Random and non-D2D UE approach in terms of SE and PC. Additionally, the insight again into the comparative evaluation of the different approaches allows one to observe that DAIS is the only approach that can react quickly to D2D Network topology changes caused through time, whether these are caused by variations in UE speed, UE direction, and number of devices in the network or TP. Additionally, our findings show that the investigated approaches achieve energy conservation and meet 5G requirements, as shown in the mMTC and eMBB use cases, even in a dynamic environment. Beyond that, as in the static case, DAIS outperforms the rest in terms of execution time, reduced message exchange, cluster formation, and control decision delay.

REFERENCES

1. T. Frank, K. Kraiss, and T. Kuhlen, "Comparative analysis of fuzzy art and ART-2A network clustering performance," *IEEE Transactions on Neural Networks*, vol. 9, pp. 544–559, May 1998.
2. G. Aydin Keskin, S. Ilhan, and C. Özkan, "The Fuzzy ART algorithm: A categorization method for supplier evaluation and selection," *Expert Systems with Applications*, vol. 37, pp. 1235–1240, Mar. 2010.
3. G. A. Carpenter, S. Grossberg, and D. B. Rosen, "Fuzzy ART: Fast stable learning and categorization of analog patterns by an adaptive resonance system," *Neural Networks*, vol. 4, pp. 759–771, Jan. 1991.
4. S. G. Akojwar and R. M. Patrikar, "Real time classifier for industrial wireless sensor network using neural networks with wavelet preprocessors," in *Proceedings of the IEEE International Conference on Industrial Technology*, pp. 512–517, Mumbai, Dec. 2006.
5. M. Ester, H.-P. Kriegel, J. Sander, and X. Xu, "A density-based algorithm for discovering clusters in large spatial databases with noise," in *KDD'96: Proceedings of the Second International Conference on Knowledge Discovery and Data Mining*, Portland, OR, pp. 226–231, AAAI Press, 1996.
6. M. Li, D. Meng, S. Gu, and S. Liu, "Research and improvement of DBSCAN cluster algorithm," *Proceedings - 2015 7th International Conference on Information Technology in Medicine and Education, ITME 2015*, pp. 537–540, Huangshan, Nov. 2016.
7. C. Dharni and M. Bnasal, "An improvement of DBSCAN Algorithm to analyze cluster for large datasets," *Proceedings of the 2013 IEEE International Conference in MOOC, Innovation and Technology in Education, MITE 2013*, pp. 42–46, Jaipur, Dec. 2013.

8. K. Khan, S. U. Rehman, K. Aziz, S. Fong, S. Sarasvady, and A. Vishwa, "DBSCAN: Past, present and future," *5th International Conference on the Applications of Digital Information and Web Technologies, ICADIWT 2014*, pp. 232–238, Bangalore, Feb. 2014.
9. G. Hamerly and C. Elkan, "Learning the k in k-means," in *Proceedings of the 16th International Conference on Neural Information Processing Systems, NIPS'03*, Cambridge, MA, pp. 281–288, MIT Press, 2003.
10. A. Smiti and Z. Elouedi, "WCOID-DG: An approach for case base maintenance based on weighting, clustering, outliers, internal detection and Dbsan-Gmeans," *Journal of Computer and System Sciences*, vol. 80, no. 1, pp. 27–38, 2014.
11. N. Li, M. Shepperd, and Y. Guo, "A systematic review of unsupervised learning techniques for software defect prediction," *Information and Software Technology*, vol. 122, p. 106287, Feb. 2020.
12. S. J. Roberts, C. Holmes, and D. Denison, "Minimum-entropy data clustering using reversible jump Markov chain Monte Carlo," *Lecture Notes in Computer Science (Including Subseries Lecture Notes in Artificial Intelligence and Lecture Notes in Bioinformatics)*, vol. 2130, no. 8, pp. 103–110, 2001.
13. H. Li, K. Zhang, and T. Jiang, "Minimum entropy clustering and applications to gene expression analysis," *Proceedings of 2004 IEEE Computational Systems Bioinformatics Conference, 2004 (CSB 2004).*, pp. 142–151, Stanford, CA, Aug. 2004.
14. F. Golchin and K. K. Paliwal, "Minimum-entropy clustering and its application to lossless image coding," *IEEE International Conference on Image Processing*, vol. 2, pp. 262–265, Santa Barbara, CA, Oct. 1997.
15. N. Rajatheva, I. Atzeni, E. Bjornson, A. Bourdoux, S. Buzzi, J.-B. Doré, S. Erkucuk, M. Fuentes, K. Guan, Y. Hu, X. Huang, J. Hulkkonen, J. Jornet, M. Katz, R. Nilsson, E. Panayirci, K. M. Rabie, N. Rajapaksha, M. Salehi, H. Sarieddeen, T. Svensson, O. Tervo, A. Tolli, Q. Wu, and W. Xu, "White paper on broadband connectivity in 6G," *arXiv: Signal Processing*, vol. abs/2004.14247, 2020.
16. C. Funai, C. Tapparello, and W. Heinzelman, "Enabling multi-hop ad hoc networks through WiFi direct multi-group networking," *2017 International Conference on Computing, Networking and Communications, ICNC 2017*, pp. 491–497, Silicon Valley, CA, Dec. 2017.
17. K. Ali, H. X. Nguyen, P. Shah, Q. T. Vien, and E. Ever, "D2D multi-hop relaying services towards disaster communication system," *Proceedings of the 24th International Conference on Telecommunications: Intelligence in Every Form, ICT 2017*, pp. 1–5, Limassol, May 2017.
18. A. Asadi and V. Mancuso, "WiFi direct and LTE D2D in action," *IFIP Wireless Days*, pp. 0–7, Nov. 2013.
19. S. Doumiati, H. Artail, and D. M. Gutierrez-Estevez, "A framework for LTE-A proximity-based device-to-device service registration and discovery," *Procedia Computer Science*, vol. 34, pp. 87–94, 2014.
20. R. F. Ustok, Interference alignment and cancellation in wireless communication systems. PhD thesis, Te Herenga Waka—Victoria University of Wellington, New Zealand, 2016.
21. R. Khosravi-Farsani and F. Marvasti, "Multiple access channels with cooperative encoders and channel state information," *arXiv*, vol. cs.IT, 2010.
22. O. A. Amodu and S. Member, "Relay-assisted D2D underlay cellular network analysis using stochastic geometry: Overview and future directions," *IEEE Access*, vol. 7, pp. 115023–115051, 2019.

23. J. Gui and J. Deng, "Multi-hop relay-aided underlay D2D communications for improving cellular coverage quality," *IEEE Access*, vol. 6, pp. 14318–14338, 2018.
24. H. Chen and F. Zhao, "A hybrid half-duplex/full-duplex transmission scheme in relay-aided cellular networks," *Eurasip Journal on Wireless Communications and Networking*, vol. 2017, p. 1, Dec. 2017.
25. I. Ioannou, C. Christophorou, V. Vassiliou, and A. Pitsillides, "5G D2D transmission mode selection performance & cluster limits evaluation of distributed AI and ML techniques," in *2021 IEEE International Conference on Communication, Networks and Satellite (COMNETSAT)*, pp. 70–80, Bali, 2021.
26. I. Ioannou, C. Christophorou, V. Vassiliou, and A. Pitsillides, "Performance evaluation of transmission mode selection in D2D communication," *NTMS 2021 Conference*, Paris, Jan. 2021.
27. S. Xiao, D. Feng, Y. Yuan-Wu, G. Y. Li, W. Guo, and S. Li, "Optimal mobile association in device-to-device-enabled heterogeneous networks," in *2015 IEEE 82nd Vehicular Technology Conference (VTC2015-Fall)*, pp. 1–5, Boston, MA, May 2015.
28. "TR 136 942 - V13.0.0 - LTE; Evolved Universal Terrestrial Radio Access (E-UTRA); Radio Frequency (RF) system scenarios (3GPP TR 36.942 version 13.0.0 Release 13)," *ETSI*, vol. 0, pp. 0–84, 2016.
29. B. Station and E. Spectrum, "LTE (FDD) transmitter characteristics," *National Telecommunications and Information Administration, United States Department of Commerce*, pp. 3–7, 2003.
30. "WiFi Direct: The worldwide network of companies that brings you Wi-Fi." [Online]. Available at: https://www.wi-fi.org/, Accessed on: 2020-09-19.
31. Qualcomm, "LTE direct; the case for device-to-device proximate discovery," *Qualcomm*, 2013.
32. R. Ware, M. K. Mukerjee, S. Seshan, and J. Sherry, "Beyond jain's fairness index: Setting the bar for the deployment of congestion control algorithms," in *Proceedings of the 18th ACM Workshop on Hot Topics in Networks, HotNets '19*, New York, NY, pp. 17–24, Association for Computing Machinery, 2019.
33. R. Jain, D. Chiu, and W. Hawe, "A quantitative measure Of fairness and discrimination for resource allocation in shared computer systems," *arXiv*, vol. abs/cs/9809099, 1998.
34. R. A. Raghuvir, D. Rajan, and M. D. Srinath, "Capacity and fairness of the multiple access channel in energy harvesting wireless networks," *International Journal of Advances in Engineering Sciences and Applied Mathematics*, vol. 5, no. 1, pp. 21–31, 2013.
35. M. Dianati, X. Shen, and S. Naik, "A new fairness index for radio resource allocation in wireless networks," in *IEEE Wireless Communications and Networking Conference, WCNC*, vol. 2, pp. 712–717, Mar. 2005.
36. H. Shi, Fairness and resource allocation in device-to-device wireless regional area network. PhD thesis, TU Delft, 2014.
37. P. Chowdhury and I. S. Misra, "A fair and efficient packet scheduling scheme for IEEE 802.16 Broadband Wireless Access Systems," *International Journal of Ad hoc, Sensor & Ubiquitous Computing*, vol. 1, no. 3, pp. 93–104, 2010.
38. T. Hofeld, L. Skorin-Kapov, P. E. Heegaard, and M. Varela, "Definition of QoE fairness in shared systems," *IEEE Communications Letters*, vol. 21, no. 1, pp. 184–187, 2017.
39. T. Hofeld, L. Skorin-Kapov, P. E. Heegaard, and M. Varela, "A new QoE fairness index for QoE management," *Quality and User Experience*, vol. 3, no. 1, pp. 1–23, 2018.

40. "5G Applications and Use Cases — Digi International." [Online]. Available at: https://www.digi.com/blog/post/5g-applications-and-use-cases, Accessed on: 2021-07-24.
41. I. F. Akyildiz, S. Nie, S. C. Lin, and M. Chandrasekaran, "5G roadmap: 10 key enabling technologies," *Computer Networks*, vol. 106, pp. 17–48, 2016.
42. D. A. Basnayaka and T. Ratnarajah, "Doppler effect assisted wireless communication for interference mitigation," *IEEE Transactions on Communications*, vol. 67, pp. 5203–5212, Jul. 2019.
43. H. Zhang, L. Venturino, N. Prasad, P. Li, S. Rangarajan, and X. Wang, "Weighted sum-rate maximization in multi-cell networks via coordinated scheduling and discrete power control," *IEEE Journal on Selected Areas in Communications*, vol. 29, pp. 1214–1224, Jun. 2011.
44. I. Ioannou, V. Vassiliou, C. Christophorou, and A. Pitsillides, "Distributed artificial intelligence solution for D2D communication in 5G networks," *IEEE Systems Journal*, vol. 14, pp. 4232–4241, Sep. 2020.
45. Mathworks, "MATLAB Mathworks MATLAB & Simulink." [Online]., 2016. Available at: https://www.mathworks.com, Accessed on: 2020-09-19.
46. "Jade Site — Java Agent DEvelopment Framework." [Online]. Available at: https://jade.tilab.com/, Accessed on: 2021-02-15.
47. L. Braubach, A. Pokahr, and W. Lamersdorf, "Jadex: A BDI-agent system combining middleware and reasoning," in *Software Agent-Based Applications, Platforms and Development Kits* (R. Unland, M. Calisti, and M. Klusch, eds.), Basel: Springer Science & Business Media, pp. 143–168, 2005.
48. A. Pokahr, L. Braubach, and W. Lamersdorf, *Jadex: A BDI Reasoning Engine*. Boston, MA: Springer US, pp. 149–174, 2005.
49. M. Ughetti, "*Jade Android Add-on Guide*," CSIE tw (http://www.csie.ncu.edu.tw), pp. 1–19, 2010.
50. G. Iavarone, T. Italia, M. Izzo, T. Italia, K. Heffner, and P. Simulation, "*Jade Tutorial Jade Programming for Android Creating the Android Chat client Project*," JADE Tilab, pp. 1–20, Jun. 2012.
51. I. Nunes, C. J. P. de Lucena, and M. Luck, "BDI4JADE: A BDI layer on top of JADE," *International Workshop on Programming Multi-Agent Systems*, pp. 88–103, Jan. 2011.
52. U. N. Kar and D. K. Sanyal, "Experimental analysis of device-to-device communication," *2019 12th International Conference on Contemporary Computing, IC3 2019*, pp. 1–6, Noida, Aug. 2019.

Notes

[1]In the Wi-Fi Direct protocol, bands are shared using multiple access channels, which reduces the interference, as shown in the [16]. For the LTE Direct, the D2D device that wants to connect with sharing device utilizes the initial orthogonal frequency that was assigned by the BS to itself in order to achieve the connection link [17–19].

[2]Demonstrating the density and how close to the means the results are spread.

[3]The spectral efficiency of the running instance divided by the total sum of the D2D devices and UEs.

[4]JFI is considered to be the standard measure of network fairness and more specifically for the QoS [32].

[5]This time is measured in each running instance and starts when a D2D device is requesting to enter in the D2D network until the transmission mode is selected and it is ready to communicate.

[6]Note that the non-D2D UE approach was used as a reference point for comparison in terms of SE and PC with the rest of the investigated approaches.

[7]Best position is where the approach forms clusters and gives the best maximum SE/minimum PC.

[8]Twenty percent for small (≤200 Devices) number of devices and 35% for large (>200) number of devices.
[9]Not less than the number of members justifying the creation of the cluster [25] neither more than the cluster head can support [26].
[10]This is accumulated to the data rate of sending a mpeg-4 HD video over a network in the minimum data rate perspective and not streaming where other factors (e.g., time, low latency) are involved (4 Mbps).
[11]Here we used the following scale to qualitative characterize the efficiency of each approach: Excellent, Very Good, Good, Average and Poor.
[12]Compared to the non-D2D UE approach, which is the QoE fairest, followed by FuzzyART, DR and DAIS.
[13]The total sum rate is the aggregated Data Rate of all links
[14]The Weighted Bandwidth rate can be calculated as a constant ratio that indicates the rate between the bandwidth of the chosen UE technology (i.e., Wi-Fi Direct, LTE Direct) and the bandwidth of the direct link towards BS.
[15]For example, with 90–100 mW TP for 50 UEs; with 130–140 mW TP for 200 UEs; and with 110–120 mW TP for 500 and 1000 UEs.
[16]Here we used the following scale to qualitative characterize the efficiency of each approach: Excellent, Good, Average and Poor.

8 Future Directions and Application Areas

In this chapter, we provide concluding remarks on this book contribution as well as summarize work in progress which extends this book in areas beyond its current scope. Furthermore, we also outline some ideas for further future work. A final concluding remark on this book is also provided at the end of this chapter.

8.1 CONCLUDING REMARKS ON BOOK CONTRIBUTION

Given the challenges and complexities of 5G and 6G, this book promotes the idea of using distributed AI (DAI) for more effective control and mobile communication. A DAI framework is designed and implemented with the realization and usage of the BDIx (extended Belief-Desire-Intention) agent in each UE. As demonstrated, this framework is expandable and can use any other technology in the BDIx agent, as for example AI/ML approaches (e.g., Generative Adversarial Network, Deep Neural Network, etc.). Additionally, the proposed framework is extensible and modular, dynamic, and adaptable, and it can monitor raised events through its sensors and architectural components, supported by reinforcement learning (RL). The RL can update the agent with the latest environment status as well as its Beliefs and AI/ML models accordingly. Also, the framework is efficient, distributed, and autonomous. This makes it resilient to existing technologies used at the Base Station. Furthermore, the framework is light in terms of resource utilization, and it can be ported and run in latest mobile devices. Additionally, it is flexible because the operator can change the value of its components and, most specifically, its Plan Library Fuzzy Logic IF-THEN Rules, any time with the use of APIs, e.g. to satisfy customer needs.

To illustrate the realization of the DAI framework and the BDIx agents, D2D is adopted as a showcase. Several plans and intentions on the use of the DAI framework are outlined in Chapter 5 to demonstrate its generality. Furthermore, to embed the concept further, the specific problem of D2D mode selection is expanded to include dynamic thresholds, from problem description to solution, and finally its evaluation to comparatively show improved mobile network SE and PC, among other performance metrics.

To demonstrate the potentials of this framework, in this work, we additionally focus on D2D transmission mode selection in 5G and develop, enhance and show DAIS (proposed in Chapter 6). DAIS is a specific plan executed by the BDIx agents for selecting the D2D Transmission mode that the D2D devices will operate, focusing on the local environment of D2D communication, rather than the global environment. Additionally, to compare DAIS with DSR, a scheme with global knowledge, we also develop and enhance it by changing the implemented algorithm and introducing the

DOI: 10.1201/9781003469209-8

same thresholds as in DAIS. Furthermore we select a number of unsupervised clustering techniques (i.e., Fuzzy ART, DBSCAN, G-MEANS and MEC) and comparatively evaluate their performance against a number of metrics (i.e. SE, PC, TP,D2D Effectiveness, D2D Stability, D2D Productivity, and QoE and QoS fairness metrics), as well as the signalling overhead and control delay in responding to changes. In the performance evaluation we include scenarios with a small and a large number of UEs, ranging from 10 to 1000.

The insight gained into the performance of enhanced DAIS and DSR, allows one to tradeoff the performance gain in terms of SE and PC versus the signalling overhead and control delay in responding to changes. Enhanced DSR performs better in terms of SE and PC, but as a distributed approach based on global knowledge, necessitates additional signalling overhead resulting in delayed control decisions. On the other hand, DAIS, which relies only on local knowledge, operates with reduced signalling overhead and much faster control decision updates, whilst remaining within 15% of the enhanced DSR performance. In addition, it was observed that the TP adjustment of the D2D devices affects in a smaller rate (<12%) the SE and affects in a high rate (>60%) the PC for all investigated approaches. Also, in terms of the three new metrics introduced, both the enhanced DAIS and DSR approaches are shown to be D2D SE effective, D2D SE stable and D2D productive.

Finally, we examine the extended DAIS approach with other competitive approaches in a dynamic environment. The results obtained demonstrated superior performance of DAIS over the SHRA, extended DSR approach, Distributed Random and non-D2D UE approach in terms of SE and PC. Additionally, the insight again into the comparative evaluation of the different approaches allows one to observe that DAIS is the only approach that can react quickly to D2D Network topology changes caused through time, either these are caused by variations in UE Speed, UE Direction, number of Devices in the network or TP. As in the static case, DAIS again outperforms the rest in terms of execution time, reduced message exchange, cluster formation, and control decision delay.

8.2 WORK IN PROGRESS AND BOOK EXTENSIONS

The material presented in this section can be considered as work in progress; its inclusion demonstrates the potential of the DAI framework and its extendability in diverse areas.

The range of topics we present include: (i) guidelines on how the DAI framework can be applied within the currently discussed standardized 5G/6G architecture. A vision on the implementation of the BDIx agents-based DAI framework D2D communication within the 5G architecture is introduced; (ii) designing a secure protocol to provide the BDIx agents the flexibility to communicate among them in a secure way; (iii) the implementation of UE-VBS using the DAI Framework approach; and (iv) an examination of DAI framework to achieve efficient Routing in D2D communication.

8.2.1 VISION ON THE IMPLEMENTATION OF THE BDIX AGENTS-BASED DAI FRAMEWORK D2D COMMUNICATION WITHIN THE 5G ARCHITECTURE

This section introduces our vision of implementing the BDIx agents based DAI framework for a D2D 5G architecture, and the realization of the agents within the software defined networking (SDN) and network function virtualization (NFV) paradigms. Below, we outline some of these constituent 5G architecture modules and discuss how we envision the DAI framework could fit within these. Note that in our architecture we do not use small cells, because our aim is to reduce small cells due to the constant PC and the all time occupied link towards the base station. The small cells are replaced with D2D-Relay devices. Our architecture is based on a small-cell architecture shown in Ref. [1].

8.2.1.1 5G Architecture and Network Slicing

Given the latest trends of network softwarization, the 5G System (5GS) architecture is composed of (i) end-to-end (E2E) network slicing; (ii) service-based architecture; (iii) software-defined networking (SDN); and (iv) network functions virtualization (NFV) [1]. With network slicing, an operator can execute various logical network instances (i.e., mobile telecom operator instances of Mobile Virtual Network (MVN) Operators) on a cooperative network infrastructure by doing constant reconciliation based on the provided Service Level Agreements (SLAs). Additionally, the lifecycle management of the network slices have to be aligned with the customer SLA. The achievement of the lifecycle management is accomplished by utilizing the service creations and service operations components from E2E frameworks. More specifically, the service level is accomplished with the closed-loop service assurance, service fulfilment, and service orchestration functions from management of domain resources NFV and Multi-access Edge Computing (MEC) with the aim to achieve orchestration throughout the lifecycle phases of the following operations: (i) preparation phase; (ii) instantiation, configuration and activation phase; (iii) run-time phase and decommissioning phase. Thus, for orchestration, closed-loop procedures are implemented to achieve the realization of the following components in the section of management of domain resources: resource fulfilment, resource assurance, and network intelligence. The components described above consist of the building blocks within each management domain. The closed-loop procedures that consist of orchestration technologies are the virtualization of network functions, software-defined programmable network functions, and infrastructure resources. Also, the SDN controllers can be programmed to efficiently execute policies and rules on the resources and functional Level. In 5G, system entities can access data from all levels as a common platform because it uses a versatile data exposure authority and access control mechanisms. The authority and mechanisms aim to provide services for data acquisition, processing, abstraction, and distribution of data related to: (i) subscribers; (ii) to the network and underlying resources; (iii) to network slice and service instances; and (iv) applications.

8.2.1.2 DAI Framework and 5G Architecture

The D2D Relays, D2D multi-hop Relays devices, and Base Station can act similar to small cells in a 5G architecture environment as shown in Ref. [1]. The proposed architecture can be supplemented according to the DAI framework functionality and framework components in our vision. Also, in our framework, the small cells under the Base Station are substituted by the D2DSHR/D2DMHR Devices. Additionally, all the Devices/components are intercommunicating with the use of API Services. Even though our framework does not need changes in the existing architecture or the 5G architecture to run, the Telecom operator needs to know how the network acts even at the edges. Because all the control and decisions are taken from the D2D Device without any other dependencies or to force control guidelines, it is necessary to monitor the D2D Devices at the edge. In addition, in a case of emergency, there are times that the network operator wants to have a predefined backhauling with ultra reliable low latency time and specific bandwidth thresholds achievement (i.e. ambulance with live video broadcasting to hospital). However, in order for the operator to monitor and measure the quality of service using the DAI Framework, the connection of the BDIx agent's actions and current state (beliefs values, Desires and current Intentions) must be logged and tracked by the architecture. In this way, in a case of emergency the operator could force control in a part of the existing D2D communication network.

In this section, we investigate how the DAI Framework and BDIx agents can be integrated into the 5G architecture. Note that in Ref. [1], the 5G architecture allows mobile core functions to be deployed close or at the mobile edge. Therefore, the service delivery in proximity to the final users is enabled. Also, in our approach, the services can be provided at the mobile edge or even let the users provide services in proximity to other users. Current virtualization technologies use a two-level virtualized execution environment. They occur in the edge data center (which resides at a location, geographically near a cluster of BSs), which allows the provision of Multi-access edge computing (MEC) capabilities to the mobile operators, improving the user experience and the dexterity in the service provisioning and delivery. Our approach utilizes both two-level virtualized environments. The only difference is that the DAI framework (BDIx agents) enabled D2D Devices will not use the distributed RRM and SON (Self-Organizing Networks) components because they are independent and autonomous. In addition, in our approach, for the same reason, the Software-Defined Radio Access Network (cD2D-SD-RAN) controller has reduced responsibilities. Therefore, as shown in Figure 8.1, the first level is the Light BS D2D Data Centre, facilitated within the Cloud-Enabled Base & D2D Devices (CE˙BS&D2Ds), which supports the execution of the Virtual Network Functions (VNFs) making up the D2D Devices access. The Light BS D2D Data Centre is envisioned to host network functions supporting traffic interception, GPRS Tunneling Protocol (GTP) encapsulation/decapsulation. The network Functional Application Platform Interface (nFAPI) can realize the connection between the D2D Devices Physical Network Functions (PNFs) and the D2D Devices VNFs. Finally, backhaul and fronthaul transmission resources will be part of the CE˙BS&D2D, allowing for the required connectivity. For the second level, as in Ref. [1], the main component of the architecture is the Main Data Centre (see Figure 8.1). The purpose of the data center is for the computation-intensive tasks and processes that need to be centralized to have a global view of the underlying infrastructure. However, most of the tasks are executed through the DAI framework at the D2D Devices, which share bandwidth; therefore, the data center responsibilities are reduced. Nevertheless, there are cases of emergency where a telecom operator should instruct a specific desire (from DAI framework) to be an intention and start implementing a specific plan, with the maximum priority to the BDIx agent due to an unexpected situation. The communication between the control data center and BDIx agent is realized using APIs from the part of the control center towards the BDIx agent and bilateral. More specifically, the BDIx agent offers API services to be accessed from the operator. However, the operator also offers API services to be accessible by the agents for bilateral communication. The cases that the operator can force beliefs (to become intentions) at BDIx agent include: (i) emergency situations (e.g. ambulance need more bandwidth and therefore the network must change in favour of the BDIx agent that is in the ambulance which, for example, will broadcast a video to a doctor); (ii) police emergency usage of bandwidth (e.g. chasing suspect); (iii) in a case of terrorist attack, army bandwidth usage; and (iv) in a case of fire, where the firefighters need to investigate the existing damage.

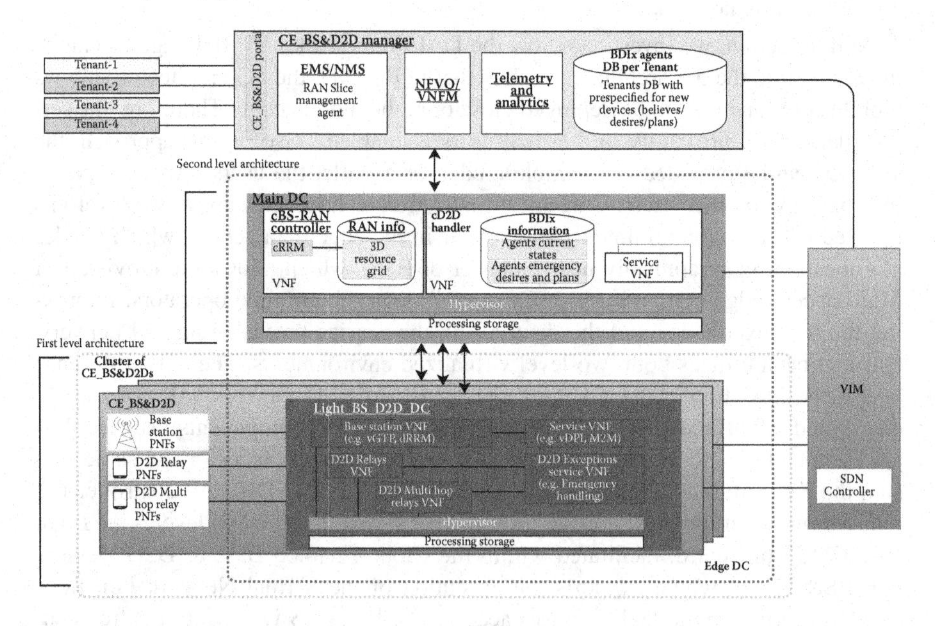

Figure 8.1 5G DAI framework architecture. (Adapted from Ref. [1].)

8.2.1.2.1 First Level Architecture

The following components are the first level important architectural components of the provided architecture: the Light BS D2D Data Centre, the Cloud Enabled Base Station, and the D2D Device.

The Light BS D2D Data Centre is integrated into the architecture of the first level. It consists of the Base Stations VNFs and provides multiple S1 (or Iu-h interface) connections from the physical Base Station (BS) to different operators' Evolved Packet Core (EPC) network elements (e.g., Mobility Management Entity (MME), SGW). Furthermore, the BS is the termination of multiple S1 interfaces connecting the CE˙BS&D2D to multiple EPC network elements (e.g. MME, SGW) entities as in S1-Flex, targeting the support of multiple tenants/operators by a single antenna. The interconnection of many BSs forms a 'cluster' that can facilitate access to a broader geographical area targeting the extension of the range while maintaining the required dexterity to provide these extensions on demand. The Light BS D2D Data Centre, also consists of two D2D Devices components: (i) The first component is for the D2D Relays VNFs representing the running BDIx agent on the D2D Device and selecting D2D Relay transmission mode; and (ii) The second represents the running BDIx agent on the D2D Device and selects D2D Multi-Hop Relays transmission mode. Additionally, it consists of the "D2D Exceptions Service VNF", the component responsible for handling emergencies in the D2D Communication Network (e.g. a fire).

In our scope, a Cloud Enabled Base Station (CE˙BS&D2D) consists of a multi-Radio Access Technology (RAT) 5G Base Station with its standard backhaul interface, standard management connection, and necessary modifications to the data model to allow Multi-Operator Core Network (MOCN) radio resource sharing. The Base Stations can have standard management connections and alterations required to the data model to enable MOCN radio resource sharing. Therefore, the Base Stations of CE˙BS&D2Ds can act as an access point (neutral BS) for network operators or virtual network operators that want to share resources at the edge of the mobile network. In addition, the BS provides to the multiple tenants/operators a Platform as a Service (PaaS) product. This service provides the deployed physical infrastructure shared among multiple network operators through BSs. Different VNFs for each BS can be hosted in the environment for different tenants. Also, the BS of CE˙BS&D2D is the termination point of the GTP-User Plane (GTP-U) tunnelling protocol, which encapsulates user IP packets from the core network entities, such as the Evolved Packet Core (EPC) Serving Gateway (SGW) in LTE, destined to the UE and vice versa. The BS of CE˙BS&D2Ds is the handling of the Radio Resource allocation in each cell that is responsible. Therefore, a module exists in the Light BS D2D Data Centre that arranges the RRA of each BS. Additionally, the CE˙BS&D2D consists of the D2D Devices selected to share their bandwidth (D2D-Relays). The representation of the sharing Devices is essential in the case of an emergency. This is the reason that D2D Exceptions Service VNF exists.

The D2D Device with the integrated BDIx agent implements the DAI framework that can support multi-Radio Access Technology (RAT). Most mobile devices have two interfaces (one Wi-Fi and one Mobile). The D2D device can support Wi-Fi Direct sharing at the Wi-Fi interface and act as a Wi-Fi client to Wi-Fi Gateway. Additionally, it can support creating backhauling links with other D2D devices using LTE Direct or connect to BS at the mobile interface. In our case, the backhauling is done among the two mobile interfaces to other D2D Devices using Wi-Fi Direct. The detection of nearby D2D-Relays (BDIx agents that act as D2D R/D2D MHR) can be achieved using Proximity Services in the Device Discovery phase, and it does not depend on the backhaul architecture. However, in terms of telecom initialization of the BDIx agent and when a life is in danger or an emergency, the D2D Devices can be monitored and controlled. Therefore, when a D2D Device enters a D2D communication network and decides its transmission mode (D2D Client, D2D-Relay), if the transmission mode is D2D-Relay it will inform the Control Centre of the decisions by calling specific REpresentational State Transfer (REST) calls to REST Application Programming Interfaces (API) services at Main Control Centre. Afterwards, the main control center will create the VNFs and associate the correct VNFs with the PNFs. When anything changes at D2D-Relay devices, the D2D Devices must inform the main control center of the change and then the main control center will release the appropriate resources.

8.2.1.2.2 Second Level Architecture

The Main Data Centre is in the second level of our proposed architecture. It encompasses the cBS-RAN controller, which is implemented as a VNF running in it. The controller makes control plane decisions for the purpose to arrange the flow of different tenant flow to specific BSs (and from the BS, the BS will forward to the destination D2D Devices) in the geographical area of the CE˙BS&D2D cluster, including the centralized Radio Resource Management (cRRM) for BSs over the entire CE˙BS&D2D cluster. Additionally, it performs cRRM decisions for handling efficiently the heterogeneous access network environment (5G RAN, LTE and Wi-Fi). These radio access networks can be programmable and are under the supervision of the centralized controller. The cBS-RAN controller updates and maintains the global network state.

In addition, our architecture can utilize other VNFs (i.e. security applications, traffic engineering, mobility management, and in general, any additional network End-to-End (E2E) service) that could be hosted by the Main Data Centre and can be deployed and managed on the virtual networks, effectively and on-demand.

Moreover, the Main Data Centre contains the cD2D Handler, which is also implemented as a running VNF, and it is responsible for handling the initialization and the setup of a new UE Device with a BDIx agent, setting up a secure D2D communications protocol and receiving API calls from the D2D Devices for monitoring. In addition, in the main DC the cD2D Handler is responsible for LTE Direct proximity messages initialization, setup and broadcasting within each BS. For this reason, there is a module that handles the LTE proximity services among each tenant called LTE ProSe Module. This module is responsible for the setup and the utilization of the proximity services of each tenant's BS provided to the dynamic selected D2D-Relays under each tenants' BS. Furthermore, the Main Data Center will execute different BSs and service VNFs under the Cloud-Enabled Base Station and D2D Manager (CE˙BS&D2DM). The CE˙BS&D2D exposes different views of the network resources: per-tenant BS and D2D Devices (D2D-Relays) view, and physical D2D Devices substrate through the BS that is managed by the network operator, decoupling the management of the BS cells from the platform itself.

8.2.1.2.3 First and Second Level Architecture

The Edge data centre (Main Data Centre and Light BS D2D Data Centre component is in both, the first and second level architecture. The Edge data centre combines the MEC and NFV concepts with D2D Device virtualization and BS Virtualization in 5G networks. In order to provide cloud services over the network infrastructure and handle the BSs as virtual resources to gather their information. The hardware modules within the architecture of the edge data center will be delivered as resources using virtualization techniques. Furthermore, combining the Edge data center architecture with the concepts of NFV and SDN will make it possible to accomplish higher levels of adaptability and versatility among the BSs.

8.2.2 SECURE COMMUNICATION PROTOCOL

In this section, we define and implement a representative secure D2D protocol that is instantiated after the execution of a secure algorithm (Plan) when the device enters the D2D communication network. More specifically, with the utilization of Digital Signatures from a well-known CA, of the device IMEI, of the Subscriber Identity Module (SIM) MSISDN/Integrated Circuit Card Identification Number (ICCID) and the time-stamped messages, the proposed security algorithm protects from fake identity, Man In The Middle Attacks, Re-transmission Attack and several other attacks. It is worth mentioning that the proposed security algorithm is executed at the application layer of ISO/OSI of the D2D communication network. Moreover, the protocol is tested and shown that is secure with the use of the Scyther tool.

8.2.2.1 The Need of a D2D Security Protocol

An open issues for D2D communication is the security aspect (see Section 2.2). The hardening of the security for D2D communications, is challenged by the following unique characteristics: (i) D2D devices establish a link among them; (ii) in our approaches (e.g., DAIS, DSR), there is a message exchange and the proposal of actions to other devices; (iii) a D2D link share involves a trust relationship among the devices; (iv) D2D Devices in order to access the gateway and internet they need to use the IP Protocol; and (v) the D2D message exchange relies on the IP protocol that is vulnerable. Overall, not much literature exists. For example there is a lack of:

- An approach that implements a light protocol of D2D communication in 5G.
- An approach that utilizes the hardware characteristics of the mobile phone (e.g., IMEI).
- An approach that uses the sim characteristics (e.g., IMSI, MSISDN) provided by the UE operators .
- An approach that utilizes the SIM storage (e.g., to save a private key in the SIM or save signature data at the SIM) from the operator at the UE .

More specifically, 4G and 5G are IP-based (Internet protocol) and heavily depend on the Internet Protocol (IP) for all the intercommunication of UEs. The BS knows all the UEs' IP addresses to communicate with each UE using the IP protocol under its cell coverage. Therefore, in any centralized control approach, the BS sends also the IP information of each D2D device with the D2D structure to the requested device. However, in the distributed control approaches such as DAIS, the entering D2D device learns about the IPs of the D2D-Relay through the ProSe messages that they send, with the utilization of LTE proximity services (e.g. see Chapter 6). Afterwards, the entering device can send communication messages to join a cluster to the "to be notified" member of the network via IP (learned from LTE proximity services) or to inform about an existing connection that will be changed and propose to the members of the altered segment of the D2D communication network to change their transmission mode. So in D2D communications, all devices have to know how to connect to some critical point devices (i.e., D2DSHR, D2DMHR) and interchange messages because of clustering and back-hauling creation.

Additionally, in the case of a raised event (e.g. the device has entered the D2D communications network), all approaches require communication with other D2D Devices or the BS in order to establish D2D communication. For example, the BDIx Agent, after DAIS executes, decides to request a change of the transmission mode of a specific D2D-Relay Device. So, the agent requests in the form of a message from the BDIx agent of the specific D2D-Relay device to change transmission mode (see Algorithm 8.1). In conclusion, security is an essential concept for the cases described above.

8.2.2.2 The D2D Security Protocol

In order to harden the D2D communications in terms of security, we implement a protocol[1] which is followed by all D2D Devices. For the protocol to run, each D2D Device and SecureProtocolServer[2] must have its own digital signatures (and know its private/public key) issued from a well recognized Certificate Authority (CA). The sign/check process is shown in Figure 8.2. Additionally,in our approach we utilize the International Mobile Equipment Identity (IMEI) number that is registered and unique in each phone along with the International Mobile Subscriber Identity (IMSI) and Mobile Station International Subscriber Directory Number (MSISDN) numbers that the UE stores in its SIM, provided by the operator. The Plan of the protocol runs in the event of "UE Enters/Leaves the D2D Network". Thus when a device enters the D2D communication network, the Desire "Security Monitoring at D2D device"

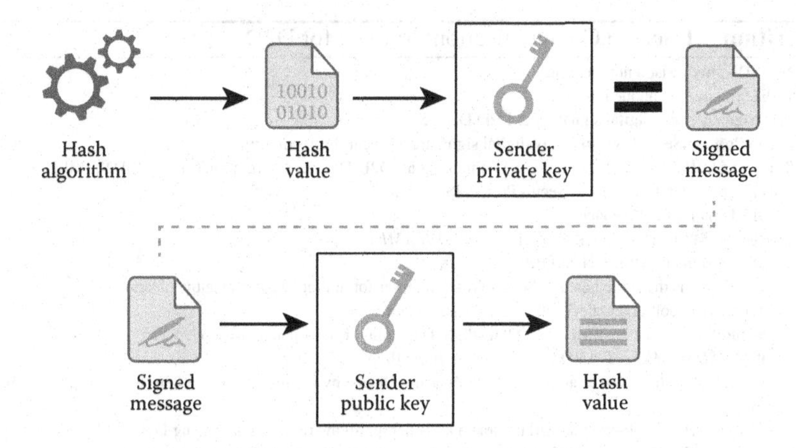

Figure 8.2 Digital signature process (public key infrastructure). (Adapted from Ref. [2].)

that always runs as Intention (as shown in Chapter 4), will run the provided Plan for establishing the protocol. Please note that after the execution of the Plan, when a device needs to send a message to another device or BS, it must include in the message the SecureProtocolServer token, and it will need to sign the message.

The Plan of the protocol shown in Algorithm 8.1 works as follows: (i) the entering D2D device signs its MSISDN and IMEI and sends them to the SecureProtocolServer for authorization and authentication; (ii) the SecureProtocolServer checks the entering D2D device signed data and verifies its signature; (iii) the SecureProtocolServer authorizes the D2D Entering Device by issuing one time token (timestamped) from Tokenizer and it signs its encryption using the entering device's public key; (iv) the SecureProtocolServer saves the token information with timestamp and D2D IMEI in the T set (a set of Data in SecureProtocolServer containing all D2D Devices information) for reference and then it sends back the encrypted signed token to the D2D Device; (v) entering D2D device decrypts the token using its private key and verifies the SecureProtocolServer digital signature; (vi) when the entering D2D device wants to send a message to another device d; (vi) the entering device creates a message for the selected D2D device "d" that with change its Transmission mode and/or its CH, SecureProtocolServer token is included and signs the message and sends to d; (vii) the d checks validity of the messages from the SecureProtocolServer. Afterwards, in any communication and message exchange the entering D2D device must include in the message the SecureProtocolServer token and sign the message that will send. Therefore, with the use of the proposed secure algorithm (shown in Figure 8.2), secure communication can be established.

BDIx agents on D2D Devices can utilize the algorithm described above in order to Authenticate/Authorized and prove identity. The DAI framework can use the protocol as it can utilize any SecureProtocolServer that has access to the telecom database holding the information of the D2D Devices (i.e. MSISDN, IMEI, IP). Also, the algorithm described above can protect from fake identity, Man In The Middle Attacks, Re-transmission Attacks and several other attacks.

Algorithm 8.1 Secure Communication Protocol for D2D

1: MSISDN: my msisdn number and
2: IMEI : my mobile phone imei code
3: D2DSignature: my digital signature at D2D Device
4: SecureProtocolServerSignature: my digital signature at SecureProtocolServer
5: T: a set of Data in SecureProtocolServer containing all D2D Devices information (i.e. MSISDN, IMEI, IP)
6: Tokenizer: Generate Tokens as SecureProtocolServer
7: D2DD: D2D Entering Device
8: **procedure** SECURITYCHECKSIND2D($T, MSISDN, IMEI$)
9:     D2DD signs its MSISDN and IMEI
10:     D2DD sends the signed data to SecureProtocolServer for authorization and authentication
11:     SecureProtocolServer checks the D2DD signed data
12:     SecureProtocolServer verify the D2DD digital signature, if is issued to a known device
13:     **if** $MSISDN, IMEI \ni T$ *AND D2DSignature is ok* **then**
14:         SecureProtocolServer authorize D2D Entering Device by issuing one time token (timestamped) from Tokenizer
15:         SecureProtocolServer signed the token and encrypt all the resulting data using D2D public key
16:         SecureProtocolServer save token information with timestamp and D2D IMEI in the T set
17:         SecureProtocolServer send the encrypted signed token to D2D Device
18:         D2DD decrypts the token using its private key
19:         D2DD verify the SecureProtocolServer digital signature, if is issued to a known device
20:         **if** *SecureProtocolServerSignature AND DATA are ok* **then**
21:             D2D entering Device compute transmission mode and proposed changes.
22:             D2D entering device generate a set of D with the affected devices (i.e. D2D, Transmission mode, CH, IP)
23:             **for each** $d \in \mathscr{D}$ **do**
24:                 D2DD creates a message for d with changes in Transmission mode and/or CH, SecureProtocolServer token is included
25:                 D2DD signs the message and sends to *d*
26:                 *d* checks token with SecureProtocolServer and signature of D2DD
27:                 **if** *D2DSignature AND token are ok* **then**
28:                     SecureProtocolServer informs new status of *d* in the *T*
29:                     d evaluates and assigns Transmission Mode ordered/requested from D2DD and/or CH
30:                 **else**
31:                     D2DD Stay connected to SecureProtocolServer
32:                 **end if**
33:             **end for**
34:         **else**
35:             D2DD Stay connected to SecureProtocolServer
36:         **end if**
37:     **else**
38:         D2DD Stay connected to SecureProtocolServer
39:     **end if**
40: **end procedure**

8.2.2.3 Experimental Results Using Scyther

The proposed protocol was confirmed for its versatility towards various attacks in the Scyther tool. Scyther is a verification tool used for the security analysis of a protocol. We assume that all functions of cryptography are perfect. The adversary cannot derive any information from the message unless he knows the decryption key. The tool is used to detect problems that arise in a given protocol and investigates if the protocol can be proven to be secure of well-known attacks. The tool is used to demonstrate security threats to the outlined Security Protocol Description Language (SPDL). The Scyther evaluates the protocol against predefined security claims which

Scyther results : verify					
Claim				**Status**	**Comments**
Secure_d2d	UEA	Secure_d2d,UEA1	Secret D2DTKA	ok	No attacks within bounc
		Secure_d2d,UEA2	Alive	ok	No attacks within bounc
		Secure_d2d,UEA3	Weakagree	ok	No attacks within bounc
		Secure_d2d,UEA4	Niagree	ok	No attacks within bounc
		Secure_d2d,UEA5	Nisynch	ok	No attacks within bounc
	BSA	Secure_d2d,UEA1	Secret ECCa	ok	No attacks within bounc
		Secure_d2d,UEA2	Alive	ok	No attacks within bounc
		Secure_d2d,UEA3	Weakagree	ok	No attacks within bounc
		Secure_d2d,UEA4	Niagree	ok	No attacks within bounc
		Secure_d2d,UEA5	Nisynch	ok	No attacks within bounc
	UEB	Secure_d2d,UEA1	Secret ECCa	ok	No attacks within bounc
		Secure_d2d,UEA2	Alive	ok	No attacks within bounc
		Secure_d2d,UEA3	Weakagree	ok	No attacks within bounc
		Secure_d2d,UEA4	Niagree	ok	No attacks within bounc
		Secure_d2d,UEA5	Nisynch	ok	No attacks within bounc

Done.

Figure 8.3 Verification of protocol until SecureProtocolServer validation.

are included in the model and validates the protocol for a bound/unbound number of sessions. In Figure 8.3, we show that the protocol is evaluated as secure.

Additionally, the tool has added functionality; it can also be used to "characterize" the defined roles in the protocol (i.e., UEa for User Equipment a, UEb for User Equipment b, BSa for Base Station that acts as SecureProtocolServer) for the purpose

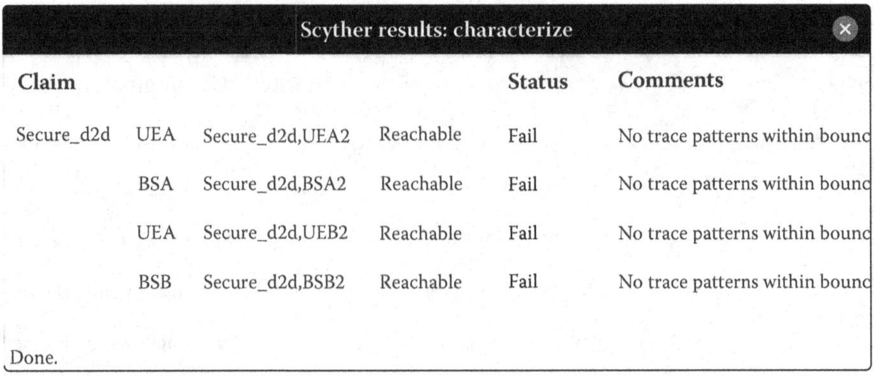

Scyther results: characterize

Claim				Status	Comments
Secure_d2d	UEA	Secure_d2d,UEA2	Reachable	Fail	No trace patterns within bound
	BSA	Secure_d2d,BSA2	Reachable	Fail	No trace patterns within bound
	UEA	Secure_d2d,UEB2	Reachable	Fail	No trace patterns within bound
	BSB	Secure_d2d,BSB2	Reachable	Fail	No trace patterns within bound

Done.

Figure 8.4 Characterization of roles.

to evaluate them as shown in Figure 8.4. Thereby performing successful execution, which demonstrates all the traces of the roles in the protocol, the status "Fail" in the figure shows no traced pattern representing an attack within the given bound.

The proposed protocol demonstrated above is shown to be secure and trustworthy. Additionally, it is shown that the DAI framework can be implemented in a secure way.

8.2.3 REALIZATION ON THE IMPLEMENTATION OF THE BDIX AGENTS-BASED DAI FRAMEWORK D2D COMMUNICATION WITHIN THE 6G ARCHITECTURE THROUGH THE ADROIT6G PROJECT

The objective of ADROIT6G is to establish a revolutionary wireless system architecture for 6G that is specifically designed to meet the performance needs of innovative and forward-thinking applications. The 5G architecture is improved by using a completely distributed and dynamic approach. This involves using distributed computing nodes at the far-edge, edge, and cloud domains. Additionally, the network is strengthened by integrating both terrestrial and non-terrestrial communications of the 6G technology. The distributed computing nodes located at the far-edge, edge, and cloud domains are utilized to implement virtual functions of software-defined disaggregated RAN and core networks, virtual applications, and AI agents. These components are dynamically orchestrated as part of the network's overall control and management strategies. The ADROIT6G architecture comprises four frameworks that operate on a programmable inter-computing and inter-network infrastructure. These frameworks include the AI-driven Management and Orchestration Framework, the Fully distributed and secure AI/ML Framework for CrowdSourcing AI Framework, the BDI- & AI-driven Unified & Open Control Operations Framework, and a Closed-Loop Functions Framework. These four frameworks collaborate to establish a complex and adaptable 6G network ecosystem capable of attaining the high goals and key performance indicators (KPIs) specified for 6G, guaranteeing the development of a future-ready 6G network landscape [3, 4]. The implementation of ADROIT6G BDIxagents can be found in the following GitHub URL: https://github.com/ADROIT6G/BDIxAgents.git (It will be public after the project completion).

8.2.3.1 The ADROIT6G Architecture

ADROIT6G (as shown in Figure 8.5) seeks to establish the groundwork for long-term research on low Technology Readiness Level (TRL) [11] technological improvements, to define a revolutionary sixth-generation (6G) wireless system architecture. The proposed architecture is designed to meet future applications' performance needs and associated technological trends. It seamlessly incorporates 6G terrestrial and non-terrestrial communications into a robust network, aiming to provide ultra-low latency support for a massive number of devices. The ADROIT6G network's architecture represents an advancement over the current 5G architecture. It adopts a fully distributed and dynamic paradigm, with distributed computing nodes (functional elements) across the far-edge, edge, and cloud domains. The distributed computing nodes, each of them with their own characteristics and capabilities, are used to deploy on demand in cloud-native environments across far-edge, edge, and cloud domains operated by different stakeholders, virtual functions of software-defined disaggregated RAN and core network, virtual applications, as well as AI agents, which are orchestrated dynamically as part of the overall network control and management strategies [3].

The ADROIT6G architecture consists of four frameworks operating on top of a programmable inter-computing and inter-network infrastructure.

The **AI-driven Management and Orchestration Framework**, which is essential for the management of the distributed applications and services across the

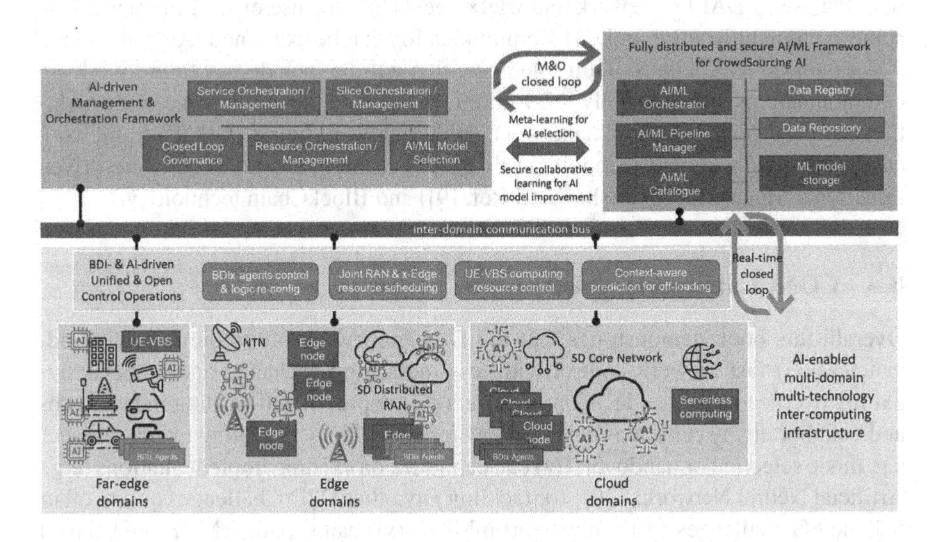

Figure 8.5 ADROIT6G's network architecture.

various infrastructures, extending from traditional cloud environments to the heterogeneous edge and the far-edge/device layer.

The **Fully distributed and secure AI/ML Framework for CrowdSourcing AI**, which represents a novel approach targeting to enhance network operations, network management, and service delivery through the use of Artificial Intelligence (AI).

The **BDI- & AI-driven Unified & Open Control Operations Framework**, which targets to harness storage, computational and networking resources of the ubiquitous smart Mobile Devices (i.e., Smartphones, Tablets, etc.), for augmenting the network's cellular, storage, and computational capabilities.

The **Closed-Loop Functions Framework**, which is designed to address the complexities inherent in managing 6G networks by leveraging automation to optimize network and service operations while efficiently utilizing infrastructure resources.

8.3 FUTURE WORK STEMMING FROM THE BOOK

Beyond the work in progress described above, future work can also include the realization of other Plans and Intentions, tackling, e.g. the rest of the D2D Challenges outlined in this book, together with extensive evaluation using both simulation and a (small scale) test-bed. Also, other challenges in 5G/6G could be tackled by using the DAI framework such as efficient routing in order to achieve the ultra reliable low latency (URLL) 5G use case, thus finally achieve all 5G uses cases. Further, a game theoretic perceptive of the BDIx agents can also be investigated, to form a multi-agent system in a non-cooperation environment, aiming to conclude in a Nash equilibrium (the game theoretic perspective of the BDIx agents that are inherited from the BDI agent). Furthermore, as additional work the implementation of UE-VBS [5–8] with the use of DAI Framework and BDIx agents and the use of DAI Framework to achieve efficient Routing in D2D Communication can be examined. Also, this book it does not directly address fault tolerance, this will be included as future work (in communications and especially D2D fault tolerance is inherent in some of the solutions, as e.g., mode selection, handover). Additionally, in future work the framework can be enriched with new technologies like D2D caching, as well as software-driven Functional Metasurfaces (as shown in Ref. [9]) and BlockChain technology.

8.4 CONCLUDING REMARK

Overall, this book demonstrates that the DAI framework offers the following advantages: (i) fast network control with less messaging exchange and reduced signalling overhead; (ii) fast decision-making; (iii) support of self-healing mechanisms and collaboratively can act as a self-organising network by executing in any disaster, e.g. mode selection or handover; (iv) can capitalize on existing implementations (e.g., Artificial Neural Networks [10]) for tackling any other D2D Challenges or any other 5G and 6G challenges; (v) can support mMTC; (vi) can support eMBB; and (vii) it can be parametrised at any time by the telecom operator.

Furthermore, this book provides different illustrative example solutions on how the DAI framework and BDIx agents can be adopted to satisfy 5G/6G challenges.

REFERENCES

1. 5G PPP Architecture Working Group, "View on 5G architecture version 3.0, June 2019," pp. 21–470, Jun 2019.
2. "Digital Signature Certificate (DSC) at Pantagon Sign Securities Pvt. Ltd.." [Online]. Available at: https://www.pantasign.com/, Accessed on: 2021-06-30.
3. C. Christophorou, et al., "ADROIT6G DAI-driven open and programmable architecture for 6G networks," *2023 IEEE Globecom Workshops (GC Wkshps)*, Kuala Lumpur, Malaysia, 2023, pp. 744-750, doi: 10.1109/GCWkshps58843.2023.10465217.
4. Patachia, C., Mihai, R., & Iordache, M., "6G initial developments for flagship applications experiments," in *2023 15th International Conference on Electronics, Computers and Artificial Intelligence (ECAI)*, pp. 1–6, IEEE, Bucharest, 2023.
5. C. Christophorou, A. Pitsillides, and I. Akyildiz, "CelEc framework for reconfigurable small cells as part of 5G ultra-dense networks," in *IEEE International Conference on Communications*, Institute of Electrical and Electronics Engineers Inc., Paris, Jul. 2017.
6. P. Swain, C. Christophorou, U. Bhattacharjee, C. M. Silva, and A. Pitsillides, "Selection of UE-based virtual small cell base stations using affinity propagation clustering," in *2018 14th International Wireless Communications and Mobile Computing Conference, IWCMC 2018*, pp. 1104–1109, Institute of Electrical and Electronics Engineers Inc., Limassol, Aug. 2018.
7. K. Venkateswararao, P. Swain, C. Christophorou, and A. Pitsillides, "Dynamic selection of virtual small base station in 5G ultra-dense network using initializing matching connection algorithm," in *International Symposium on Advanced Networks and Telecommunication Systems, ANTS*, vol. 2019-December, IEEE Computer Society, Goa, Dec. 2019.
8. K. Venkateswararao, P. Swain, C. Christophorou, and A. Pitsillides, "Using UE-VBS for dynamic virtual small cells deployment and backhauling in 5G ultra-dense networks," *Computer Networks*, vol. 189, p. 107926, Apr 2021.
9. A. Pitilakis, A. C. Tasolamprou, C. Liaskos, F. Liu, O. Tsilipakos, X. Wang, M. S. Mirmoosa, K. Kossifos, J. Georgiou, A. Pitsilides, N. V. Kantartzis, S. Ioannidis, E. N. Economou, M. Kafesaki, S. A. Tretyakov, and C. M. Soukoulis, "Software-defined metasurface paradigm: Concept, challenges, prospects," in *2018 12th International Congress on Artificial Materials for Novel Wave Phenomena (Metamaterials)*, pp. 483–485, Espoo, Aug. 2018.
10. A. M. Schweidtmann and A. Mitsos, "Deterministic global optimization with artificial neural networks embedded," *Journal of Optimization Theory and Applications*, vol. 180, no. 3, pp. 925–948, 2019.
11. Technology Readiness Levels, NASA 2023, Available online at: https://www.nasa.gov/directorates/somd/space-communications-navigation-program/technology-readiness-levels/, Accessed on: 2024-09-15.

Notes

[1]The proposed protocol forms the basis of on-going collaboration between the Computer Science Department of the University of Cyprus and Dept of Electronics And Communication Department Faculty, SSN institutions, Chennai, INDIA.

[2]BS or other authenticate cloud device that is online and has access to the operators database.

Appendix A

A.1 DAIS AND DSR COMMON TERMS AND PARAMETERS

The terms and parameters that are not only used for DAIS but also used for DSR are provided below:

D2DSHR: D2D Relay/D2D Single-hop Relay
D2DMHR: D2D Multi-hop Relay
D2DCH: D2D Cluster Head
WDR: Weighted Data Rate (Used only in DAIS)
SR: Sum Rate (Used only in DSR)
MAXUsersCH: Maximum users supported by a D2DCH = 200
MAXQueryD2DRelayDistance: Maximum distance for querying D2DSHRs = 200 m
MAXDistancetoFormCluster: Maximum distance of D2D devices from the D2DSHR acting as D2DCH for accepting connections = 200 m
MAXSpeedToFormBackhauling: Maximum speed of the D2D device in order to operate as D2D-Relay = 1.5 m/s (pedestrian)
MAXDistanceMultiHop: Maximum distance of a D2D device from the nearest D2DSHR in order to operate as D2DMHR = 1000 m
MAXDistanceMoveAway: Maximum distance that a D2D device acting as D2D Client/D2DSHR moves away from its connected D2D-R, in order to rerun the Transmission Selection Algorithm (DAIS/DSR) = 200 m
PERCDataRate: This is associated with the WDR threshold in DAIS and the LDR threshold in DSR. Its value is expressed in percentage (%) and considered by a D2D device[1] in order to: (i) decide the Transmission mode that will operate; or (ii) decide if and how the D2D network structure will alter (see also Section 6.1.3.1)
DeviceBatteryThreshold: This is associated with the BPL threshold. This threshold determines the minimum value (in percentage) that the remaining battery level of a D2D device must be, in order to be able to become a D2DSHR or a D2DMHR and accept connections from other D2D devices (see also Section 6.1.3.2).
maxD2DSHR: The D2DSHR with the maximum WDR (for DAIS) or SR (for DSR) within MAXQueryD2DRelayDistance distance from the D2D device that is running the Transmission mode Selection algorithm (DAIS or DSR). The formulas used to estimate this parameter can be found in Appendix A.2.
maxD2DMHRNoConnections:[2] The D2DMHR with the maximum WDR (for DAIS) or SR (for DSR) and with no connection links with other D2DSHRs/D2D Clients located within MAXDistancetoFormCluster distance from the D2D device that is running the transmission mode selection algorithm. The formulas used to estimate this parameter can be found in Appendix A.2.

maxD2DSHRNoConnectionsToBeD2DMHR:[3] The D2DSHR with the maximum WDR (for DAIS) or SR (for DSR) and with no connection links with other D2DSHRs/D2D Clients located within MAXDistanceMultiHop distance from the D2D device that is running the transmission mode selection algorithm. The formulas used to estimate this parameter can be found in Appendix A.2.
maxD2DSHRToUseUED2DMHR:[4] The D2DSHR with the maximum WDR (for DAIS) or SR (for DSR), but worst than one of the D2D devices that is running the transmission mode selection algorithm, and with no connection links with other D2D Clients located within MAXDistanceMultiHop distance from the D2D device. The formulas used to estimate this parameter can be found in Appendix A.2.
maxD2DMHRToUseAsMultiHop:[5] The D2DMHR with the maximum WDR (for DAIS) or SR (for DSR) and with no connection links with other D2DSHRs/D2D Clients located within MAXDistanceMultiHop distance from the D2D device that is running the transmission mode selection algorithm. The formulas used to estimate this parameter can be found in Appendix A.2.
WeightedDataRateSelectedD2DR: The link data rate among Candidate D2D and maxD2DR.
DR: The data rate among the candidate D2D Device and the BS.
DataRateThreshold: Its value is expressed in percentage (%) and considered by a D2D device[6] in order to do quality check, when a device is valuable to connect as client to the D2D relay device.
SelectedD2DR: A selected D2D relay from the D2D relays that when the Candidate D2D connects to, it achieves the maximum sum rate compared to the other D2D relays. If the D2D Candidate considers to be D2D Client, the distance constraint (MAXDistancetoFormCluster) is taken under consideration.
DataRateSelectedD2DR: The link data rate among Candidate D2D and SelectedD2DR.
SumRateIfSelectD2DClient: The sum rate of whole network plus the DataRateSelectedD2DR.
D2DRSelectedD2DMHRorBS: A selected D2D multi hop relay from the D2D multi hop relays that when the Candidate D2D connects to as D2D relay or D2D multi hop relay, it achieves the maximum sum rate compared to the other D2D multi hop relays. If the D2D Candidate considers to be D2D relay, the distance constraint (MAXQueryD2DRelayDistance) is taken under consideration. If the D2D Candidate considers to be D2D multi hop relay, the distance constraint (MAXDistanceMultiHop) is taken under consideration.
SumRateIfSelectD2DR: The sum rate of whole network plus the link among the D2D Candidate device and the D2DRSelectedD2DMHRorBS when D2D Candidate is D2D relay.
SumRateIfSelectD2DMHR: The sum rate of whole network plus the link among the D2D Candidate device and the D2DRSelectedD2DMHRorBS when D2D Candidate is D2D multi relay.

Table A.1
Algorithm Notations and Mathematical Representations of Parameters

Notations	Mathematical Representation
d	$\sqrt{(\mathrm{UE}x_1 - \mathrm{D2D}x_2)^2 + (\mathrm{UE}y_1 - \mathrm{D2D}y_2)^2}$
maxD2DSHR	$\mathrm{D2D}_j$ where $\mathrm{WDR}_{\mathrm{D2D}_j} = (\max(\mathrm{WDR}_{\mathrm{D2D}_i})\exists\ \mathrm{D2D}_i$ where $d \geq$ MAXDistancetoFormCluster $\wedge\, \mathrm{WDR}_{\mathrm{D2D}_i} \geq (\mathrm{WDR}_{\mathrm{UE}_i} + \mathrm{PERCDataRate} * \mathrm{WDR}_{\mathrm{UE}_i}) \wedge i \in \mathrm{D2DSHR}$ $\wedge$ COUNT($\mathrm{D2D}_{ig}$ where g served by $i) \leq D$)
maxD2DMHRNoConnections	$\mathrm{D2D}_j$ where $\mathrm{WDR}_{\mathrm{D2D}_j} = (\max(\mathrm{WDR}_{\mathrm{D2D}_i})\exists\ \mathrm{D2D}_i$ where $d \geq$ MAXDistancetoFormCluster$\wedge$ $\mathrm{WDR}_{\mathrm{D2D}_i} \geq (\mathrm{WDR}_{\mathrm{UE}_i} + \mathrm{PERCDataRate} * \mathrm{WDR}_{\mathrm{UE}_i}) \wedge i \in \mathrm{D2DMHR} \wedge$ COUNT($\mathrm{D2D}_{ig}$ where g served by $i) = 0$)
maxD2DSHRNoConnectionsToBeD2DMHR	$\mathrm{D2D}_j$ where $\mathrm{WDR}_{\mathrm{D2D}_j} = (\max(\mathrm{WDR}_{\mathrm{D2D}_i})\exists\ \mathrm{D2D}_i$ where $d \geq$ MAXDistancetoFormCluster$\wedge$ $d \leq$ MAXQueryD2DRelayDistance $\wedge\, \mathrm{WDR}_{\mathrm{D2D}_i} \geq (\mathrm{WDR}_{\mathrm{UE}_i} + \mathrm{PERCDataRate} * \mathrm{WDR}_{\mathrm{UE}_i}) \wedge$ $i \in \mathrm{D2DSHR} \wedge$ COUNT($D2D_{ig}$ where g served by $i) = 0) \wedge$ $\mathrm{D2DDevicePower}_i \geq$ DeviceBatteryThreshold
maxD2DSHRToUseUED2DMHR	$\mathrm{D2D}_j$ where $\mathrm{WDR}_{\mathrm{D2D}_j} = (\max(\mathrm{WDR}_{\mathrm{D2D}_i})\exists\ \mathrm{D2D}_i$ where $d \geq$ MAXDistancetoFormCluster $\wedge$ $d \leq$ MAXQueryD2DRelayDistance $\wedge\, \mathrm{WDR}_{\mathrm{D2D}_i} \ll (\mathrm{WDR}_{\mathrm{UE}_i} - \mathrm{PERCDataRate} * \mathrm{WDR}_{\mathrm{UE}_i})$ $\wedge\, i \in \mathrm{D2DSHR} \wedge \mathrm{D2DDevicePower}_i \geq$ DeviceBatteryThreshold
maxD2DMHRToUseAsMultiHop	$\mathrm{D2D}_j$ where $\mathrm{WDR}_{\mathrm{D2D}_j} = (\max(\mathrm{WDR}_{\mathrm{D2D}_i})\exists\ \mathrm{D2D}_i$ where $d \geq$ MAXQueryD2DRelayDistance$\wedge$ $d \leq$ MAXDistanceMultihop $\wedge\, \mathrm{WDR}_{\mathrm{D2D}_i} \geq (\mathrm{WDR}_{\mathrm{UE}_i} + \mathrm{PERCDataRate} * \mathrm{WDR}_{\mathrm{UE}_i}) \wedge$ $i \in \mathrm{D2DMHR} \wedge$ COUNT($\mathrm{D2D}_{ig}$ where g served by $i) = 0) \wedge$ $\mathrm{D2DDevicePower}_i \geq$ DeviceBatteryThreshold

A.2 DAIS AND DSR COMMON FORMULAS FOR PARAMETER ESTIMATION

Please note that the mathematical formulation of the above parameters and terms is shown in Table A.1.

Notes

[1] A D2D device that is running the Transmission Mode Selection algorithm (DAIS or DSR).

[2] The selected D2DMHR will change transmission mode to D2DSHR and the D2D investigated Device will connect to it as D2D Client.

[3] The selected D2DSHR will change its transmission mode to D2DMHR and the D2D device running the transmission mode selection algorithm will set its transmission mode to D2DSHR and will connect to it.

[4] The D2D device running the transmission mode selection algorithm will select the D2DMHR mode and the D2DSHR will connect to it.

[5] The D2D device running the transmission mode selection algorithm will set its transmission mode to D2DSHR and connect to the D2DMHR.

[6] A D2D device that is running the transmission mode selection algorithm (Sum Rate).

Appendix B

B.1 DISTRIBUTED ARTIFICIAL INTELLIGENCE POWER RESERVATION PLAN BASED ON TP

The distributed artificial intelligent power reservation (DAIPPR) plan (see Algorithm B.1) will be executed when the D2D-relay battery power level reduces less than a threshold (i.e., 50%; this threshold can be set by the operator). The aim is to prevent D2D-relay battery drain and lose of connections of the D2D-Clients it serves. Additionally, in order for the plan to be triggered, the D2D-relay checks first if with TP alteration, the following are met: (i) the percentage change of PC (as formulated in Eq. B.1) is more or equal with 50% and (ii) the percentage change of SE (as formulated in Eq. B.2) is less or equal with 15%. It is important to highlight here that these values are selected empirically by considering extensive simulation and the results provided in Table 7.6.

$$\min \mathrm{PC}_{\mathrm{tp}}(\mathrm{UEs}, \mathrm{app}) = \min_{x=60,\ldots,160} (f_{\mathrm{pc}}(\mathrm{UEs}, x, \mathrm{app})) \tag{B.1}$$

$$\min \mathrm{SE}_{\mathrm{tp}}(\mathrm{UEs}, \mathrm{app}) = \min_{x=60,\ldots,160} (f_{\mathrm{se}}(\mathrm{UEs}, x, \mathrm{app})) \tag{B.2}$$

$$G(\mathrm{UEs}, \mathrm{tra}_{\mathrm{power}}, \mathrm{app}) = \frac{f_{\mathrm{pc}}(\mathrm{UEs}, \mathrm{tra}_{\mathrm{power}}, \mathrm{app}) - \min \mathrm{PC}_{\mathrm{tp}}(\mathrm{UEs}, \mathrm{app})}{f_{\mathrm{pc}}(\mathrm{UEs}, \mathrm{tra}_{\mathrm{power}}, \mathrm{app})} \times 100 \tag{B.3}$$

$$T(\mathrm{UEs}, \mathrm{tra}_{\mathrm{power}}, \mathrm{app}) = \frac{f_{\mathrm{se}}(\mathrm{UEs}, \mathrm{tra}_{\mathrm{power}}, \mathrm{app}) - \min \mathrm{SE}_{\mathrm{tp}}(\mathrm{app})}{f_{\mathrm{se}}(\mathrm{UEs}, \mathrm{tra}_{\mathrm{power}}, \mathrm{app})} \times 100 \tag{B.4}$$

Algorithm B.1 Distributed Artificial Intelligent Power Reservation (DAIPR) Algorithm for reducing Transmission Power

```
 1: D2DSHR˙D2DMHR: The D2DSHR or D2DMHR Device
 2: BatteryPower: Battery Power Level
 3: D2D˙Clients: The number of D2D Clients that D2DSHR˙D2DMHR serves
 4: TP: The Transmission Power of the communication link between the D2DSHR˙D2DMHR and the BS/D2DMHR
 5: TP˙acceptable˙min: Minimum acceptable TP
 6: StatiStics˙PC: Array with the Transmission Power change and PC change % for DAIS for 0 - 1000 D2D devices
 7: StatiStics˙SE: Array with the Transmission Power change and SE change % for DAIS for 0 - 1000 D2D devices
 8: NumberOFUEs: The total number of D2D devices UEs under our D2D communication Network taken by LTE
    ProSe
 9: procedure PowerReservationAlgorithm(StatiStics˙PC_NumberOFUEs,
          StatiStics˙SE(NumberOFUEs), BatteryPower, TP,
          BatteryPower, D2D˙Clients)
10:     if BatteryPower ≦ 50 % ∧ D2D˙Clients ≧ 1 then
11:         Set Operation˙TP with TP˙acceptable˙min
12:         for all elements ∈ StatiStics˙PC_NumberOFUEs do
13:             Set percentage˙change as percentage change of investigated element
14:             Set Investicated˙TP as the Transmission Power of investigated element
15:             if (percentage˙change ≧ 50%) then
16:                 Set percChangeofPC_Investicated˙TP = percentage˙change
17:             end if
18:         end for
19:         for all elements ∈ StatiStics˙SE_NumberOFUEs do
20:             Set percentage˙change as percentage change of investigated element
21:             Set Investicated˙TP as the Transmission Power of investigated element
22:             if (percentage˙change ≦ 15%) then
23:                 Set percChangeofSE_Investicated˙TP = percentage˙change
24:             end if
25:         end for
26:         if ∃(count(percChangeofPC) ≧ 1 ∧ count(percChangeofSE) ≧ 1) then
27:             Sort elements of percChangeofPC in descending order base of the Perc. Ch.
28:             Sort elements of percChangeofSE in increasing order base of the Perc. Ch.
29:             set Found false
30:             for all get tp from Perc. Ch. element ∈ percChangeofPC do
31:                 for all get tp2 from Perc. Ch. element2 ∈ percChangeofSE do
32:                     if ∃(tp ≡ tp2 ∧ tp ≠ TP) then
33:                         SetOperation˙TP tp
34:                         Set Found true
35:                         break
36:                     end if
37:                 end for
38:                 if (Found) then
39:                     break
40:                 end if
41:             end for
42:         end if
43:     end if
44: end procedure
```

Index

Printed in the United States
by Baker & Taylor Publisher Services

Printed in the United States
by Baker & Taylor Publisher Services